PARKINSON'S DISEASE — II

Aging and Neuroendocrine Relationships

ADVANCES IN EXPERIMENTAL MEDICINE AND BIOLOGY

Recent Volumes in this Series

Volume 105
NUTRITIONAL IMPROVEMENT OF FOOD PROTEINS
Edited by Mendel Friedman

Volume 106
GASTROINTESTINAL HORMONES AND PATHOLOGY OF THE DIGESTIVE SYSTEM
Edited by Morton Grossman, V. Speranza, N. Basso, and E. Lezoche

Volume 107
SECRETORY IMMUNITY AND INFECTION
Edited by Jerry R. McGhee, Jiri Mestecky, and James L. Babb

Volume 108
AGING AND BIOLOGICAL RHYTHMS
Edited by Harvey V. Samis, Jr. and Salvatore Capobianco

Volume 109
DRUGS, LIPID METABOLISM, AND ATHEROSCLEROSIS
Edited by David Kritchevsky, Rodolfo Paoletti, and William L. Holmes

Volume 110
HUMAN INTERFERON: Production and Clinical Use
Edited by Warren R. Stinebring and Paul J. Chapple

Volume 111
HORMONES AND ENERGY METABOLISM
Edited by David M. Klachko, Ralph R. Anderson, and Murray Heimberg

Volume 112
OVARIAN FOLLICULAR AND CORPUS LUTEUM FUNCTION
Edited by Cornelia P. Channing, John M. Marsh, and William A. Sadler

Volume 113
PARKINSON'S DISEASE – II: Aging and Neuroendocrine Relationships
Edited by Caleb E. Finch, David E. Potter, and Alexander D. Kenny

Volume 114
FUNCTION AND STRUCTURE OF THE IMMUNE SYSTEM
Edited by Wolfgang Müller-Ruchholtz and Hans Konrad Müller-Hermelink

Volume 115
DYNAMICS OF ARTERIAL FLOW
Edited by Stewart Wolf and Nicholas T. Werthessen

PARKINSON'S DISEASE—II
Aging and Neuroendocrine Relationships

Edited by

Caleb E. Finch
University of Southern California
Los Angeles, California

and

David E. Potter and
Alexander D. Kenny
Texas Tech University School of Medicine
Lubbock, Texas

PLENUM PRESS • NEW YORK AND LONDON

Library of Congress Cataloging in Publication Data

Tarbox Parkinson's Disease Symposium, 2d, Lubbock, Tex., 1978.
Parkinson's disease – II.

(Advances in experimental medicine and biology; v. 113)
Includes index.
1. Brain chemistry – Congresses. 2. Neuroendocrinology – Congresses. 3. Aging – Congresses. 4. Parkinsonism – Congresses. I. Finch, Caleb Ellicott. II. Potter, David Edward, 1937– III. Kenny, Alexander D. IV. Texas Tech University. Dept. of Pharmacology and Therapeutics. V. Tarbox Parkinson's Disease Institute. VI. Texas Tech University. Office of Continuing Education and Faculty Development. VII. Title. VIII. Series.
QP376.T37 1978 616.8'33 79-491

ISBN 978-1-4684-8895-1 ISBN 978-1-4684-8893-7 (eBook)
DOI 10.1007/978-1-4684-8893-7

Proceedings of the Second Tarbox Parkinson's Disease Symposium held at Lubbock, Texas, February 2–4, 1978, and sponsored by

Texas Tech University School of Medicine
Department of Pharmacology and Therapeutics
Tarbox Parkinson's Disease Institute
Office of Continuing Education and Faculty Development

Softcover reprint of the hardcover 1st edition 1978

A Division of Plenum Publishing Corporation
227 West 17th Street, New York, N.Y. 10011

SYMPOSIUM CHAIRMAN

ALEXANDER D. KENNY, Ph.D.
Professor and Chairman
Department of Pharmacology and Therapeutics
Director, Tarbox Parkinson's Disease Institute
Texas Tech University School of Medicine
Lubbock, Texas, U.S.A.

SESSION CHAIRMEN

CHARLES D. BARNES, Ph.D.
Professor and Chairman
Department of Physiology
Texas Tech University School of Medicine

JACK DUNN, JR., M.D.
Professor and Chief
Division of Neurologic Surgery
Department of Surgery
Texas Tech University School of Medicine

J. BARRY LOMBARDINI, Ph.D.
Associate Professor
Department of Pharmacology and Therapeutics
Texas Tech University School of Medicine

WILLIAM F. MAURER, M.D.
Assistant Professor
Department of Pediatrics
Texas Tech University School of Medicine

PETER K.T. PANG, Ph.D.
Associate Professor
Department of Pharmacology and Therapeutics
Texas Tech University School of Medicine

JAMES H. PIRCH, Ph.D.
Associate Professor
Department of Pharmacology and Therapeutics
Texas Tech University School of Medicine

DAVID E. POTTER, Ph.D.
Associate Professor
Department of Pharmacology and Therapeutics
Texas Tech University School of Medicine

THOMAS E. TENNER, JR., Ph.D.
Assistant Professor
Department of Pharmacology and Therapeutics
Texas Tech University School of Medicine

INVITED SPEAKERS

AVRID CARLSSON, M.D.
Department of Pharmacology
University of Göteborg
Göteborg, Sweden

CALEB E. FINCH, Ph.D.
Department of Molecular Biology and Andrus Gerontology Center
University of Southern California
Los Angeles, California, U.S.A.

PATRICK L. McGEER, M.D., Ph.D.
Department of Neurology and Kinsmen Laboratory of Neurological Research
Faculty of Medicine
University of British Columbia
Vancouver, British Columbia
Canada

RICHARD C. ADELMAN, Ph.D.
Department of Biochemistry
Director, Temple University Institute on Aging
Temple University School of Medicine
Philadelphia, Pennsylvania, U.S.A.

JAMES A. CLEMENS, Ph.D.
Lilly Research Laboratories
Eli Lilly and Company
Indianapolis, Indiana, U.S.A.

ANDRÉ BARBEAU, M.D.
Department of Neurobiology
Clinical Research Institute of Montreal
Montreal, Quebec, Canada

OTHER CONTRIBUTORS

RICHARD F. WALKER, Ph.D.
Center for the Study of Aging and Human Development
Duke University Medical Center
Durham, North Carolina, U.S.A.

MAHLON M. WILKES, Ph.D.
Department of Reproductive Medicine
School of Medicine
University of California at San Diego
La Jolla, California, U.S.A.

GARY D. GRAY, Ph.D.
Department of Physiology
Stanford University School of Medicine
Stanford, California, U.S.A.

GAIL D. RIEGLE, Ph.D.
Endocrine Research Unit
Department of Physiology
Michigan State University
East Lansing, Michigan, U.S.A.

PHILIP W. LANDFIELD, Ph.D.
Department of Physiology and Pharmacology
Bowman Gray School of Medicine
Winston-Salem, North Carolina
U.S.A.

YVONNE C. CLEMENT-CORMIER, Ph.D.
Departments of Pharmacology and Neurobiology
University of Texas Medical School at Houston
Houston, Texas, U.S.A.

MAYNARD H. MAKMAN, Ph.D.
Department of Biochemistry
Albert Einstein College of
of Medicine
Yeshiva University
New York, New York, U.S.A.

JOHN R. SLADEK, JR., Ph.D.
Department of Anatomy
University of Rochester School
of Medicine and Dentistry
Rochester, New York, U.S.A.

PETER DAVIES, Ph.D.
Department of Pathology
Albert Einstein College of
Medicine
Yeshiva University
New York, New York, U.S.A.

ACKNOWLEDGMENTS

The Second Tarbox Parkinson's Disease Symposium was supported in part by grants from:

Merck, Sharp and Dohme Postgraduate Program
Endo Laboratories, Inc.
DHEW Grant 5 D08 PE 16072

PREFACE

These Proceedings emanate from the Second Tarbox Parkinson's Disease Symposium held February 2-4, 1978, at the SouthPark Inn in Lubbock, Texas. The Symposium was sponsored by the Tarbox Parkinson's Disease Institute and the Department of Pharmacology and Therapeutics of the Texas Tech University School of Medicine at Lubbock. The First Symposium took place in October, 1976.

The Second Tarbox Parkinson's Disease Symposium boldly brought together investigators at the cutting edge of aging and neuroendocrine research and attempted to relate them to Parkinson's disease. Credit for the concept must go to the Program Chairman, Dr. David E. Potter. Once the plan was conceived the organizers relied heavily on the advice of Dr. Caleb E. Finch, whose counsel in developing the program was invaluable. The final verdict on the success of this daring venture must await the outcome of the publication of this Volume. Nevertheless, those in attendance could not fail but note the enthusiasm and excitement of the participants as their findings from the diverse disciplines of the neurological, aging, and endocrine sciences converged. This in itself is testimony that the Symposium accomplished some degree of success.

The Tarbox Parkinson's Disease Institute was established in 1973 with funds appropriated by the State of Texas and is dedicated to research, patient care, and education in Parkinson's disease and related neurological disorders. The Institute is named after Mr. Elmer L. Tarbox, who served the Lubbock area as a Representative in the Texas Legislature, and is himself a victim of Parkinson's disease.

The Tarbox Fellowship Programs were instituted in 1977 and represent the current research thrust of the Tarbox Institute. These include the Tarbox Postdoctoral Fellowships, the Tarbox Predoctoral Research Fellowships, and the Tarbox Medical Student Research Fellowships. The Institute also conducts a regularly scheduled Tarbox Clinic for Parkinson's Disease at the Texas Tech University School of Medicine in Lubbock.

Thanks are especially due to Virginia M. Davis and her office staff of the Department of Pharmacology and Therapeutics for the production of camera-ready copy for the Publisher. In particular, the dedication of Cheri L. Mathewson, who bore the brunt of the responsibility for the typing and the uniformity of format of the final product, is deeply appreciated.

I wish to dedicate this Volume to the late Maxine Tarbox, whose untimely death in July, 1978, was a shock to all who knew and loved her. Mrs. Elmer L. Tarbox had worked untiringly and unceasingly in support of the Tarbox Parkinson's Disease Institute. Her vibrant and friendly personality will be greatly missed.

Alexander D. Kenny, Ph.D.

CONTENTS

AGE-DEPENDENT CHANGES IN CENTRAL DOPAMINERGIC AND OTHER MONOAMINERGIC SYSTEMS

ARVID CARLSSON

Department of Pharmacology
University of Göteborg, Fack
S-400 33 Göteborg, Sweden

This presentation will focus on three apparently critical stages during the life-span of central dopaminergic neurons: (1) the vulnerable period of differentiation in pre- and early postnatal life; (2) the changes at puberty, induced by the sexual hormones; and (3) the decline of brain dopamine during adult life. Other monoaminergic neurons will also be discussed for the purpose of comparison.

It is now generally accepted that central dopaminergic neurons play an important role for at least three different aspects of brain function: (1) the control of motor functions exerted by the extrapyramidal system; (2) the regulation of some fundamental psychic processes, such as alertness, initiative and thought organization; and (3) the control of endocrine functions, such as the secretion of prolactin by the anterior pituitary. The main dopaminergic pathways apparently involved in these functions are, respectively, the nigrostriatal, the mesolimbic and the tubero-infundibular pathways (*for review and references see Carlsson, 1977*).

THE VULNERABILITY OF DOPAMINERGIC NEURONS DURING EARLY POSTNATAL LIFE

The thalidomide catastrophe taught us that chemical agents may interfere specifically with ontogenic processes without necessarily being harmful to mature cells. Critical and vulnerable phases of cell differentiation are not confined to the early pregnancy but may, in fact, extend to early postnatal life. Nerve cells are known to reach full maturity at a rather late stage, and different types of nerve cells vary in their rate of maturation. Already in 1961, Bertler suggested, on the basis of observations on human fetal brains, that dopamine-storing structures develop at a later stage than noradrenaline-storing structures in brain. This has later been confirmed *(Agrawal and Himwich, 1970; Connor and Neff, 1970; Loizou, 1972)*. Moreover, functional and biochemical studies *(see Lundborg and Kellogg, 1973)* indicate that central dopamine-receptor mechanisms in rats and rabbits develop during the first few postnatal weeks and clearly more slowly than the noradrenaline-receptor mechanisms.

In order to test the hypothesis that receptors are specifically involved in the development of synapses, Lundborg *(1972)* administered small doses of haloperidol, a dopamine receptor-blocking agent, to rabbit mothers, in a dose of 1 mg/kg daily via the drinking water for the first seven days after parturition. Control mothers were given the drug vehicle. This treatment had no detectable effect on the gross behavior of the mothers or on the growth of the young. However, at the age of 8 days the young of the haloperidol-treated mothers displayed considerable gait problems when placed on a smooth surface outside the cage. They showed a marked inability to raise their bodies and heads. This disturbance was still evident, though less pronounced, one week later. Even at the age of 4 weeks a muscular dysfunction of the hindlimbs was detectable. It should be noted that a single huge dose of haloperidol to the young did not induce this disturbance. These observations led to the hypothesis that exposure of the young to haloperidol continuously during the first week of life leads to blockade of central dopamine receptors, and that this blockade interferes with the development of dopaminergic synapses.

This hypothesis was further tested in rats, using two other dopamine-receptor antagonists. Nursing rat mothers were given pimozide 0.5 mg/kg intraperitoneally daily the first seven days after delivery, or penfluridol 1 mg/kg/orally of day 1,3 and 5 after delivery, or the appropriate vehicles. Both mothers and young showed normal gross behavior throughout the study and there was no difference in growth between penfluridol- and vehicle-treated animals. At the age of 4 weeks the animals were trained to a conditioned avoidance response in a two-way shuttle box. The acquisition of this response was considerably retarded in the experimental as compared to control rats (Figure 1). The brains of the experimental animals showed abnormalities in monoamine, especially dopamine, turnover. In particular, the synthesis of dopamine in the limbic regions was markedly retarded (Figure 2). That these biochemical changes were relevant for the behavioral disturbance was supported by the observation that amphetamine, a catecholamine-releasing agent, was capable of restoring the response to normal *(for review see Lundborg and Engel, 1975)*. In a subsequent investigation *(Ahlenius, Engel, Hård, Larsson, Lundborg and Sinnerstedt, 1977)* the open-field behavior or the offspring of nursing rat mothers given penfluridol as described above was found to be abnormal, with a higher than normal ambulation at 4 weeks and the opposite change at 8 and 12 weeks of age. The pattern of behavorial changes resembles the clinical so-called minimal brain dysfunction

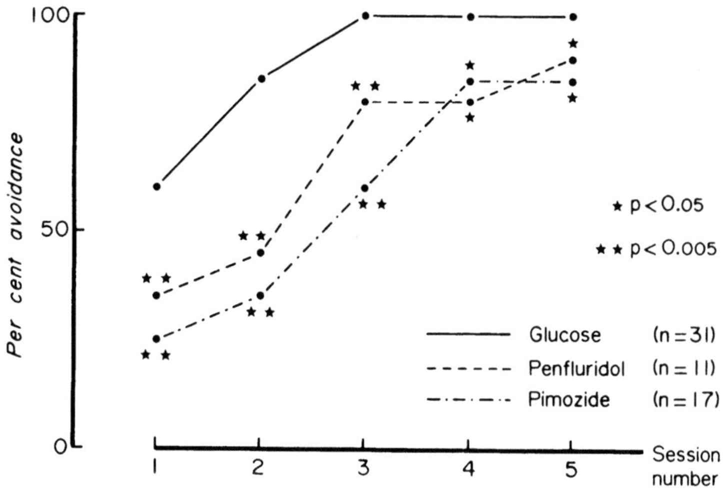

Figure 1. *Acquisition of a conditioned avoidance response (CAR) by offspring of rat mothers treated with dopamine-receptor antagonists during the first week after delivery (modified from Lundborg and Engel, 1975).*

Nursing rat mothers were given: (a) pimozide (0.5 mg/kg i.p.) or 5.5% glucose (2 mg/kg at day 1,2,3, 4,5,6 and 7 after delivery; or (b) penfluridol (1 mg/kg p.o.) or 5.5% glucose (5 ml/kg at day 1,3,and 5 after delivery). The offspring were trained in a shuttle box 4 weeks after birth in five consecutive daily sessions. Per cent avoidance values are medians (N=17 and 16, respectively). The p values refer to comparison between experimental and control groups in pimozide series (Mann-Whitney U-test).

syndrome observed in children. This syndrome is sucessfully treated with amphetamine.

These observations suggest that interference with a transmitter-receptor mechanism at a critical stage of maturation will cause a disturbance in the development of the synaptic function which is probably permanent.

If receptor activation plays an essential role for the development of synapses in general, some important consequences may be considered. Such dependence on receptor function may help to explain for example, the mental retardation in phenylketonuria. Here the high plasma level of phenylalanine leads to reduced brain uptake of other large neutral amino acids, e.g. tyrosine and tryptophan, and the lowered brain levels of these amino acids *(McKean, 1972)* probably cause reduced synthesis of catecholamines and 5-HT *(Carlsson and Lindqvist, 1978)*, and this, in turn, may lead to

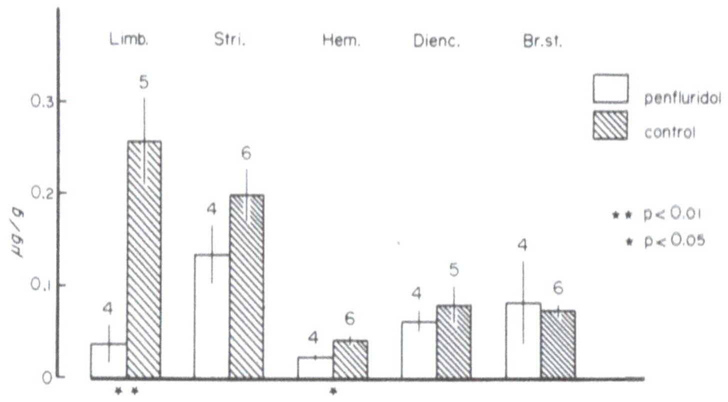

*Figure 2. Accumulation of DOPA after inhibition of the aromatic L-aminoacid decarboxylase by NSD 1015 (100 mg/kg i.p. 30 min before death) in various brain regions of 28-day-old offspring to penfluridol-treated controls (from Lundborg and Engel, 1975). The treatment schedule was the same as in Fig. 1. Limb: limbic forebrain (dopamine-rich structures); Stri: striatum; Hem: rest of cerebral hemispheres (containing predominantly noradrenaline); Dienc: diencephalon; Br st: lower brain stem. Numbers above bars are the number of (pools) brain-part samples. Shown are the means ± SE; *p<0.05; **p<0.01.*

reduced activation of dopamine, noradrenaline and 5HT receptors, followed by failure of the respective synapses to develop. Similarly, the mental retardation in cretinism may be at least partly due to failure of noradrenaline synapses to develop, since the sensitivity of central α-adrenergic receptors has been shown to depend on the thyroid state *(See Strömbom, Svensson, Jackson and Engström, 1977)*.

The critical period of nerve cell differentiation may be assumed to start during pregnancy and to last for one or several years postnatally.

CASTRATION-INDUCED CHANGES IN BRAIN-MONOAMINE SYNTHESIS

Puberty represents another critical period in the development of central nervous function. In the rat, puberty occurs at the age of 50 to 60 days. Our studies are so far confined to male rats. We have so far found no striking change in the synthesis of brain monoamines at puberty, except for a moderate decrease in 5-HT synthesis. This finding, in conjunction with the castration-induced increase in 5-HT synthesis reported below, would fit in with the alleged inhibitory action of 5-HT on sexual activity. In any event, castration reveals a clearcut influence of the gonads on brain monoamines. The formation of dopa is enhanced in the dopamine-rich limbic and striatal regions, and so is the formation of 5-HT, especially in the limbic and diencephalic regions. The regional specificity of these changes is striking. For example, in the main part of the cerebral hemispheres (excluding striatum and limbic forebrain) the effect of castration is slight or absent (see Figure 3). The onset of these changes is remarkably slow and is evident only 20 days after castration. Moreover, when castration was performed before puberty, no significant changes occured 20 days later. Substitution therapy by subcutaneous implantation of an adequate dose of testosterone prevented the observed castration effects *(Engel, Ahlenius, Almgren and Carlsson, 1978)*.

Remarkably enough, castration also caused an increase in tyrosine levels, but again only in the dopamine-rich areas where dopa formation was increased (Figure 3). Some moderate increases in tryptophan were also observed. While the mechanisms underlying these changes in amino acid levels remain obscure, they are probably not important causative factors for the observed increases in monoamine-synthesis rates. In the case of brain-tyrosine hydroxylase, it has been found to be almost fully saturated with tyrosine *(Carlsson and Lindqvist, 1978)* and thus the moderate increase in tyrosine levels observed would not be expected to cause any detectable increase in dopa formation.

At present, it is not possible to interpret these castration-induced changes in monoamine synthesis in precise functional terms. The regional selectivity and the preventive effects of substitution therapy suggest that we are dealing with specific actions mediated via hormone receptors. We did not observe any changes in tyrosine-hydroxylase activity *in vitro* and thus the increased *in vivo* activity probably occurred without any change in the number of existing enzyme molecules. Cofactor availability and a conformational change in the enzyme molecules are possible factors that remain to be considered. One intriguing possibility is that testosterone enhances the sensitivity of dopamine receptors and that the castration-induced changes are feedback mediated, analogous to the stimulation of dopamine synthesis and turnover by dopamine-receptor antagonists. The antipsychotic action of these agents is thought to be mediated via an antidopaminergic action in the limbic system. Thus, the effect of castration on dopa formation in the dopamine-rich limbic forebrain regions is of considerable interest. As a point of speculation it may be recalled that the onset of schizophrenia often occurs at puberty, or shortly afterwards, and thus an involvement of sexual hormones in the pathogenesis or schizophrenia cannot be excluded. Low plasma testosterone levels have been reported in schizophrenic patients *(Brambilla, Guerrini, Rizzi and Ricciardi, 1974)*.

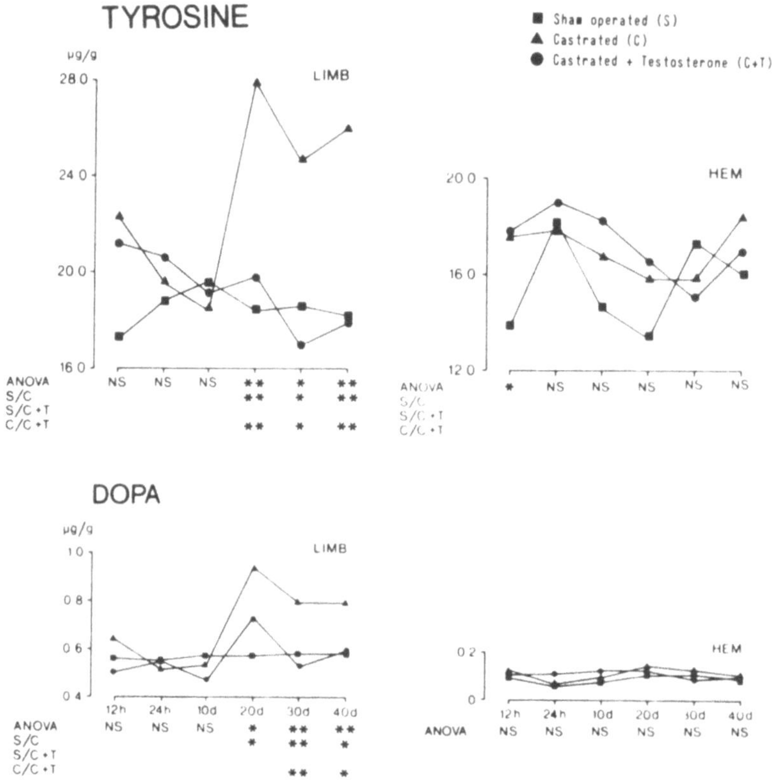

*Figure 3. Effect of castration of male rats on the accumulation of DOPA after inhibition of the aromatic L-aminoacid decarboxylase by NSD 1015 (100 mg/kg i.p. 30 min before death) and tyrosine levels in dopamine-rich limbic forebrain regions and in a noradrenaline-predominated cerebral hemisphere portion (from Engel et al., 1978). The rats were castrated (C) or sham-operated (S) at the age of 70 days and were killed at various intervals after surgery. A third group of animals was castrated and given substitution therapy by subcutaneously implanted testosterone (C + T). Shown are the means of the two pooled brain-region samples. Significance levels refer to Student's t-test after one-way analysis of variance; *$p<0.05$; **:$p<0.01$.*

The question arises why there are no more striking changes in the synthesis of brain monoamines at puberty, as might be expected from the effects of the castration, apart from the moderate decrease in 5-HT synthesis mentioned above. A possible explanation might be that the action of testosterone on monoamine synthesis is balanced by the opposite action of another factor developing at puberty.

Investigations of the influence of the pituitary on the brain monoamines may possibly help to throw light on this problem.

THE DECREASE IN BRAIN DOPAMINE WITH AGE

The loss of brain neurons with age appears to vary considerably between different brain regions and types of neurons. For example, brain-stem neurons in general are much more resistant to age-induced losses than the neurons of the cerebral cortex. However, certain brain-stem neurons, such as those of the locus coeruleus and substantia nigra, which contain noradrenaline and dopamine, respectively, form exceptions. They have been reported to be reduced in number in the human senescent brain *(Brody, 1973, 1977)*.

The dopamine content of the basal ganglia has been shown to decrease with increasing age in animals *(Finch, 1976; Algeri, Ponzio, Bonati and Brunello, 1976)* and in man (Figure 4, *Carlsson and Winblad, 1976; Adolfsson, Gottfries and Winblad, 1976)*. In mice, noradrenaline was not reduced in senescent brains *(Finch, 1976)* and in man only a slight, statistically not significant, decrease has been found *(Carlsson and Winblad, unpublished data; Gottfries, et al., unpublished data)*. Similarly, tyrosine

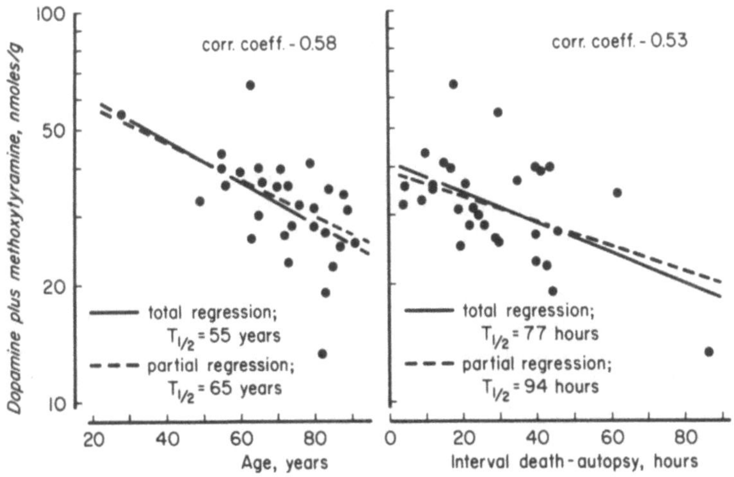

Figure 4. Dopamine (+ methoxytyramine, mainly formed from dopamine postmortem). Level in human putamen: semilogarithmic plots of regression upon age (left) and death-autopsy interval (right). $T\frac{1}{2}$ = half-life. Partial regression coefficients were obtained by a multiple regression analysis of the log amine levels versus age and time interval (from Carlsson and Winblad, 1976).

hydroxylase activity appears to be more strongly negatively correlated with age in dopamine-rich brain areas *(McGeer and McGeer, 1976)*.

It seems reasonable to assume that age-dependent neuronal degeneration is a common denominator for these various changes in dopamine neurons with respect to cell number, transmitter content and activity of synthesizing enzymes, even though the contribution by other factors cannot be excluded. The fact that corresponding changes in noradrenaline content and synthetic capacity appear to be less pronounced, underlines the selectivity of the age-induced neuronal degeneration (See Table 1).

In our studies on the dopamine content of the human brain the interval between death and autopsy was found to be a significant factor. Both dopamine and noradrenaline were negatively correlated to this interval. Part of the catecholamines was converted into the corresponding 3-O-methylated base, i.e. 3-methoxytyramine and normetanephrine, respectively. Animal data indicate that a major fraction of these metabolites found in brain has been formed postmortem, and its respective O-methylated metabolite is closer to the true catecholamine level at the time of death than the value of the catecholamine alone. However, even this sum is negatively correlated with the postmortem delay, indicating that other processes, e.g. oxidative deamination, contribute to the postmortem loss *(Carlsson, Lindqvist and Kehr, 1974)*.

After correction of the dopamine in the putamen, respectively, for age and postmortem delay, we found that the time of day when death occurs is a significant factor. Doapmine appeared to have a maximum and a minimum at about 6 PM and 6 AM, respectively (Figure 5). Circadian variations in monoamine levels have been observed in animals, too *(Reis, Weinbren and Corvelli, 1968; Manshardt and Wurtman, 1968)*. Such variations may be of interest in several respects: (1) they may throw light on the physiology of sleep and wakefulness; (2) the circadian rhythms man be specifically disturbed in pathological states, e.g. depression and senility, and contribute to their symptomatology; and (3) correction for these variations as well as those related to age and postmortem delay may facilitate the detection of changes induced by pathological states.

Our data do not yet permit us to decide whether an age-dependent loss of 5-HT occurs in human brain (see Table 1).

COMMENT

It would appear that the dopaminergic neurons of the brain lead an adventurous life. At least three critical stages can be distinguished, i.e. the perinatal period, puberty, and senescence. Each of these periods may involve risks: (1) perinatal disturbances in development and differentiation, induced by drugs and various toxic factors, infections, malnutrition, ischemia, *et cetera*, may lead to functional abnormalities, e.g. mental retardation; (2) er-

TABLE 1. STATISTICAL DATA ON REGRESSION OF LOG AMINE LEVELS IN HUMAN BRAIN REGIONS UPON AGE AND DEATH-AUTOPSY TIME INTERVAL (from Carlsson and Winblad, unpublished data).

	Dopamine + methoxy-tyramine in putamen n = 37	Noradrenaline + normeta-nephrine in hypothalamus n = 37	5-HT in hypothalamus n = 20
Age years			
Correlation coeff.	- 0.52****	- 0.21	- 0.33
Regression coeff. (total)	- 0.0038****	- 0.0038	- 0.0071
Regression coeff. (partial)	- 0.0036***	- 0.0020	- 0.0065
Time interval, hours			
Correlation coeff.	- 0.23	- 0.46***	- 0.45*
Regression coeff. (total)	- 0.0018	- 0.0089***	- 0.15*
Regression coeff. (partial)	- 0.00096	- 0.0085***	- 0.015[a]
Multiple correlation coeff.	0.54****	0.47***	0.54**

* $p<0.05$; ** $p<0.02$; ***$p<0.01$; ****$p<0.001$; a) $p \sim 0.05$.

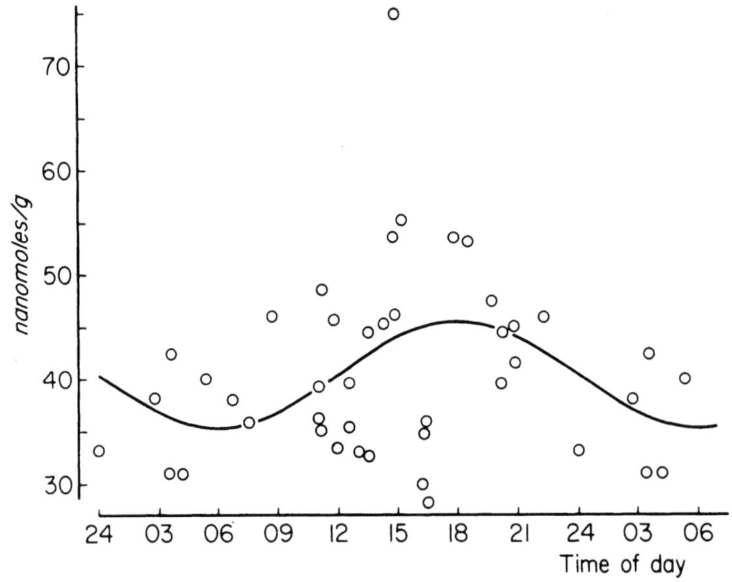

Figure 5. Dopamine (+ methoxytyramine, mainly formed from dopamine postmortem). Level in human putamen: circadian variation. Shown are single values corrected to age 50 years and death-autopsy interval zero hours.

rors in the profound reprogrammation of various brain function induced by the sex hormones at puberty may possibly play a role in the pathogenesis of schizophrenia and other behavioral abberations of adolescence; and (3) neuronal losses of old age may cause mental and motor disturbances, including Parkinson's disease, senile and presenile dementia. It should be realized that early deviations may not immediately lead to overt symptoms. For example, deficient development of dopaminergic neurons at an early stage may not *per se* be severe enough to induce functional abnormalities. However, as a consequence of age-dependent neuronal loss the deficiency may become manifest and show up, for example, as parkinsonism or presenile dementia. Parkinsonian patients appear to consist of two subgroups, one with a profile suggesting a more widespread neuronal damage with less striking response to dopa and a higher incidence of dementia *(Granérus, 1977)*. It seems that we have to consider a continuum, where one extreme is represented by a highly selective damage of the nigrostriatal dopamine pathway and the other by a rather widespread neuronal degeneration, even though degeneration of dopaminergic neurons may be represented by senile dementia. In fact, Gottfries and his colleagues have found a negative correlation between senile dementia and brain homovanillic acid *(Gottfries, Gottfries and Roos, 1978; and unpublished data)*. Also, they report that dopa therapy has a favorable influence on the dementia symptoms.

If dopamine deficiency plays a role in senile dementia, why do not patients suffering from senile dementia also have parkinsonian symptoms? The answer could be that the parkinsonian symptom picture is the result of a disturbed balance between different transmitter systems. For example, there is evidence that in this disease the cholinergic system in the striatum predominates over the deficient dopaminergic system. If the imbalance in corrected by the administration of an anticholinergic drug, the symptoms are alleviated, though not as effectively as if dopa is given. In senile dementia there is perhaps a parallel decline of acetylcholine *(Bowen, Smith, White and Davidson, 1976; Davies and Maloney, 1976; Perry, Perry, Blessed and Tomlinson, 1977; White, Goodhardt, Keet, Hiley, Carrasco and Williams, 1977)* and dopamine leading to an undisturbed striatal balance, whereas the effects on the intellectual capacity are perhaps potentiated. Therefore the combined therapy with dopa and cholinergic agents, e.g. the precursor choline, seems worthwhile. Perhaps a new field is about to be opened up, namely that of substitution therapy for brain neurotransmitter deficiency. In this context one more line of research should be mentioned. Lehmann *(1966)* in our Department has described a stuporous syndrome which is apparently due to tryptophan deficiency; treatment with tryptophan results in sometimes dramatic improvement.

Mental disturbances of old age may thus be partly due to deficiency of one or more transmitters and will perhaps in the future be treated with precursors or analogues of transmitters. Sometimes we may be dealing with deficiency of just one transmitter, but in other cases the deficiency may be multiple and may require treatment with "cocktails".

REFERENCES

Adolfsson, R., Gottfries, C.G. and Winblad, B. (1976). Methodological aspects of postmortem investigations of human brain with special reference to monoamines and enzymes. Paper presented at Xth CINP meeting, Quebec, Canada.

Agrawal, H.C. and Himwich, W.A. (1970). Aminoacids, proteins, and monoamines of developing brain. In: *Developmental neurobiology*, pp. 287-310. Ed. W.A. Himwich. Springfield: Ch. C. Thomas.

Ahlenius, S., Engel, J., Hard, E., Larsson, K., Lundborg, P. and Sinnerstedt, P. (1977). Open field behavior and gross motor development in offspring of nursing rat mothers given penfluridol. *Pharmacology, Biochemistry and Behavior 6*, 343-347.

Algeri, S., Ponzio, F., Bonati, M. and Brunello, N. (1976). Biochemical changes in monoaminergic nerves in the CNS of the senescent rat. Paper presented at Xth CINP meeting, Quebec, Canada.

Bertler, A. (1961). Occurrence and localization of catecholamines in the human brain. *Acta physiol scand. 51*, 97-107.

Bowen, D.M., Smith, C.B., White, P. and Davidson, A.N. (1976). Neurotransmitter-related enzymes and indices of hypoxia in senile dementia and other abiotrophies. *Brain 99*, 459-496.

Brambilla, F., Guerrine, A., Rizzi, F. and Ricciardi, F. (1974). Psychoendocrine investigations in schizophrenia: Relationship between pituitary-gonadal function and behavior. *Dis. Nerv. Syst. 2*, 362-367.

Brody, H. (1973). Aging of the vertebrate brain. In: *Development and aging in the nervous system*, pp. 121-133. Eds. M. Rockstein and M.L. Sussman. New York: Adac. Press.

Brody, H. (1977). Paper presented at Symposium on Aging, Institut de la Vie, Vichy, France.

Carlsson, A. (1977). Does dopamine play a role in schizophrenia? *Psychological Medicine 7*, 583-597.

Carlsson, A. and Lindqvist, M. (1978). Dependence of 5-HT and catecholamine synthesis on precursor aminoacid levels in rat brain, in preparation.

Carlsson, A., Lindqvist, M. and Kehr, W. (1974). Postmortal accumulation of 3-methoxytyramine in brain. *Nauyn-Schmiedeberg's Arch. exp. Path. Pharmak. 284*, 365-372.

Carlsson, A. and Winblad, B. (1976). Influence of age and time interval between death and autopsy on dopamine and 3-methoxytyramine levels in human basal ganglia. *J. Neural Transm. 38*, 271-276.

Connor, J.D. and Neff, N.H. (1970). Dopamine concentration in the caudate nucleus of the developing cat. *Life Sci. 9*, 1165-1168.

Davies, P. and Maloney, A.J.F. (1976). Selective loss of central cholinergic neurons in Alzheimer's disease. *Lancet ii*, 1403.

Engel, J., Ahlenius, S., Almgren, O. and Carlsson, A. (1978). Effects of gonadectomy and hormone replacement on brain-monoamine synthesis in male rats, in preparation.

Finch, C. (1976). The regulation of physiological changes during mammalian aging. *Q. Rev. Biol. 51*, 49-83.

Gottfries, C.G. (1978). Biochemical determinants of dementia. In: *Handbook of biological psychiatry. Brain mechanisms and abnormal behavior*, Vol. 2, Eds. H.M. van Praag, M.G. Lader, C.J. Rafaelsen and E.J. Sachar. New York: Marcel Dekker, Inc.

Gottfries, C.G., Gottfries, I. and Roos, B. -E. (1969). The investigation of homovanillic acid in the human brain and its correlation to senile dementia. *Br. J. Psychiat. 225*, 563-574.

Granérus, A.-K. (1977). L-Dopa treatment in Parkinson's syndrome. Thesis. Kungälv. Sweden: Gotab.

Lehmann, J. (1966). Mental disturbances followed by stupor in a patient with carcinoidosis. *Acta psychiat. scand. 42*, 153-161.

Loizou. L.A. (1972). The postnatal ontogeny of monoamine-containing neurons in the central nervous system of the albino rat. *Brain Res. 40*, 395-418.

Lundborg, P. (1972). Abnormal ontogeny in young rabbits after chronic administration of haloperidol to the nursing mothers. *Brain Res. 44*,684-687.

Lundborg, P. and Engel, J. (1975). Learning deficits and selective biochemical brain changes in 4-week old offspring of nursing rat mothers treated with neuroleptics. In: *Antipsychotic drugs: pharmacodynamics and pharmacokinetics*, pp. 261-269. Eds. G. Sedvall, B. Uvnds and Y. Zotterman. Oxford: Pergamon Press.

Lundborg, P. and Kellogg, C. (1973). Pharmacological approaches to monoamine receptors during brain devélopment. In: *Dynamics of degeneration and growth in neurons*, pp. 561-573. Eds. K. Fuxe, L. Olson and Y. Zotterman. Oxford: Pergamon Press.

Manshardt, J. and Wurtman, R.J. (1968). Daily rhythym in the noradrenaline content of rat hypothalamus. *Nature 217*, 574-575.

McGeer, P.L. and McGeer, E.G. (1976). Enzymes associated with the metabolism of catecholamines, acetylcholine, and gaba in human controls and patients with Parkinson's disease and Huntington's chorea. *J. Neurochem. 26*, 65-76.

McKean, Ch. M. (1972). The effects of high phenylalanine concentrations on serotonin and catecholamine metabolism in the human brain. *Brain Res. 47*, 469-476.

Perry, E.K., Perry, R.H.,Blessed, G. and Tomlinson, B.E. (1977). Necropsy evidence of central cholinergic deficits in senile dementia. *Lancet i*, 189.

Reis, D.J., Weinbren, M. and Corvelli, A. (1968). A circadian rhythm of norepinephrine regionally in cat brain: its relstionship to environmental lighting and to regional diurnal variations in brain serotonin. *J. Pharmac. exp. Ther. 164*. 135-145.

Strombom, U., Svensson, R.J., Jackson, C.M. and Engstrom, G. (1977). Hyperthyroidism: Specifically increased response to central NA-(α-)receptor stimulation and generally increased monoamine turnover in brain. *J. Neural Transm. 41*, 73-92.

White, P., Goodhardt, M.J., Keet, J.P., Hiley, C.R., Carrasco, L.H. and Williams, I.E.I. (1977). Neurocortical cholinergic neurons in elderly people. *Lancet i*, 668.

AGE-RELATED CHANGES IN BRAIN CATECHOLAMINES: A SYNOPSIS OF FINDINGS IN C57BL/6J MICE AND OTHER RODENT MODELS*

CALEB E. FINCH

Department of Molecular Biology and the
Andrus Gerontology Center
University of Southern California
Los Angeles, California 90007

INTRODUCTION

During the past few years it has become clear that altered functions of several neurotransmitter systems occur during normal aging in short- and long-lived mammals. As will be described below and elsewhere in this symposium *(Carlsson, Davies, Makman, McGeer, Sladek, Wilkes)* post-maturation impairments occur in dopaminergic, noradrenergic, serotonergic, and cholinergic systems. A major contemporary issue concerns the role of neuronal loss in the reported changes. The available information on the C57BL/6J male mouse will be reviewed in depth, since at present this strain is one of the best characterized rodent models for biochemical, endocrinological and pathological changes of aging.

AGING CHARACTERISTICS OF THE C57BL/6J MOUSE

The C57BL/6J is a major stock available from Jackson Laboratories, Bar Harbor, Me., where it has been inbred and maintained under close genetic observation since 1936 *(Staats, 1976)*. When fed on high protein, low fats diets such as Purina Lab Chow, male mice achieve maximum, stable adult weights of between 25 to 35 g (mean of 28g) by 6 to 8 mo *(Finch, 1969; Finch, Foster, Mirsky, 1969; Finch and Foster, 1973)*. Unless a mouse becomes sick, adult body weight varies by ≤ 3g (10% mean weight) *(Finch et al., 1969)*. In carefully maintained but conventional (i.e. not germ free) colony conditions *(Finch, 1969; Finch et al., 1969)*, mortality rate remains low until 24 mo (Figure 1); thereafter the incidence of a variety of benign and pathological lesions (described below) rises followed by a sharp rise in mortality. In general, the typical lifespan is 28

*Contribution No. 36 from the Laboratory of Neurobiology

TABLE 1. NEURONAL LOSS AND AGING (A PARTIAL SURVEY)

Region	Species	Reference	Extent of loss during the lifespan
locus coeruleus	human	Brody, 1976	20-40%
substantia nigra	human	McGeer et al., 1977	20-40%
superior central gyrus	human	Brody, 1970	20-45%
ventral cochlear nucleus	human	Konigsmark and Murphy, 1970	≤15%
abducens nucleus	human	Vijayashankar and Brody, 1971	≤15%
trochlear nucleus	human	Vijayashankar and Brody, 1973	≤15%
inferior olivary n.	human	Monagle and Brody, 1974	≤15%
spinal roots	mice	Wright and Spink, 1959	≤15%
	rats	Sant'Ambrogio et al., 1961	≤15%
	cats	Moyer and Kalizewsky, 1958	≤15%
	humans	Corbin and Gardner, 1937	≤15%

to 30 mo, with a maximum of about 44 mo *(Finch, 1971; Kunstyr and Luenberger, 1975)*. Characteristic age-related lesions of C57BL/6J mice include: greatly enlarged seminal vesicles [an almost ubiquitous finding after 24 mo, which is associated with greater volume of seminal fluid and glandular atrophy; the seminal vesicle cell population is decreased by 50% according to DNA content *(Finch and Girgis, 1974)*]; lymphoreticular disease which may involve liver, spleen or mesenteric lymph nodes *(Dunn, 1954; Hanna, Nettesheim and Snodgrass, 1971; Finch and Foster, 1973)*; reduced hematocrits *(Finch and Foster, 1973)*; or sudden weight gain or loss *(Finch et al., 1969)*. In 10 years, we have never observed gross evidence of tumors in the pituitary or other endocrine glands in the C57BL/6J male. In contrast, pituitary tumors are common in females after 24 mo *(Felicio, Nelson and Finch, unpublished observations)*.

AGING AND DISEASE

A major problem in experimental gerontology concerns the varied effect of age-related lesions, since not all individuals in older age groups will have the same lesions *(Finch, 1972 and 1977a; Finch and Foster, 1973; Nelson, Latham and Finch, 1975).* A study of testicular functions demonstrates the impact of age-related diseases: mice with major pathologic lesions such as lymphoreticular tumors, respiratory infections, or hydronephrosis also showed reduced plasma testosterone and testicular weight. In contrast, a healthy group of mice (70% of survivors aged 28 mo) selected for the absence of gross lesions had identical distributions of plasma testosterone to young mice *(Nelson et al., 1975)* and responses of testes to LH *in vitro (Finch, Jonec, Wisner, Sinha, de Vellis and Swerdloff, 1977).* The apparent maintenance of testicular endocrine functions in healthy C57BL/6J mice during aging may not apply to all strains of mice *(Bronson and Desjardins, 1977)* or rats *(e.g. Riegle and Miller, 1978).* [N.B. it is currently difficult to

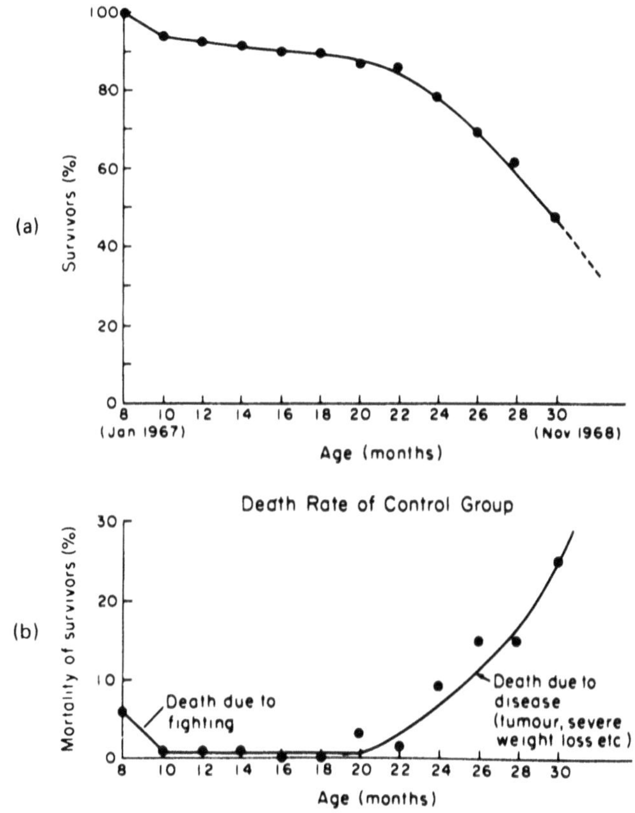

Figure 1. *(a) Survival of C57BL/6J male mice (n-96) as a function of age. The maximum longevity of this cohort was 42 mo (Finch, 1971).*
(b) Age specific mortality rats. For details of disease-related deaths, see Finch et al. (1969).

interpret most studies on aging rats because of the high incidence of tumors in the pituitary *(Durbin, Williams, Jeung and Arnold, 1966; Greisbach, 1967; Duchen and Schurr, 1976)* and testes *(Coleman, Barthold, Osbaldiston, Foster and Jonas, 1977)*]. Finally, it is possible that some of the variability in human plasma testosterone values in older men *(Vermeulen, Reubens and Verdonck, 1972)* reflects a distribution of age-related diseases, which affect some men more than others. In any case, it is now clear that there is no event in men which is comparable to the total loss of fertility and 90% reduction of ovarian estradiol and progesterone after menopause (reviewed in *Finch and Flurkey, 1977*).

NEUROTRANSMITTER LEVELS IN C57BL/6J MICE

In the following studies, comparisons are made between young, fully grown adult (8 to 12 mo) and senescent (26-30 mo) C57BL/6J mice; all mice used were free of gross lesions, except for the very common enlarged seminal vesicles *(Finch and Girgis, 1974, see above)*.

WHOLE BRAIN

Postmaturational, age-related changes have not been revealed in numerous measurements of the putative neurotransmitters, e.g. dopamine, norepinephrine, serotonin, glutamic acid, GABA and their enzymes in whole brains (reviewed in *Finch, 1977b*). This profile of apparent constancy does not contradict data (described below) showing changes localized to specific brain regions and neural systems. In the brain, as in most (if not all) organs, age-related alterations of function are selective and cell-type specific *(Finch et al., 1969; Finch, 1976)*.

BRAIN REGIONS

It is obvious that subdivision of the brain into such regions as cortex, cerebellum, hypothalamus, brain stem, etc., which are readily dissected, does not restrict the functions controlled to that region, nor does it delineate the neuronal projections involved. Nonetheless, in this very early stage of the subject, it is useful to consider the available data by brain region.

Catecholamines were measured by our adaptation of a radioenzymatic assay *(Coyle and Henry, 1973; Palkovits, Brownstein, Saavedra and Axelrod, 1974)*. After labelling catecholamines *in vitro* with ^{3}H-methyl (from S-adenosyl methionine with hepatic catechol-O-methyl transferase), the methylated derivatives of dopamine, norepinephrine and epinephrine are acetylated and separated by 1-d thin layer chromatography *(Osterburg and Finch, unpublished observations)*. This method gives us blank values of 50 to 150 cpm, $\leq$ 2% overlap between the amines, and a sensitivity (2x blank) of 2 to 5 pg in routine tissue assays.

We find substantial decreases of dopamine in the striatum

TABLE 2. AGE-RELATED CHANGES IN BRAIN CATECHOLAMINES IN C57BL/6J MICE

	ng/mg protein ± SEM		
	9-12 mo	25-28 mo	Significance
Striatum*			
dopamine	46.4 ± 3.9	32.9 ± 2.2	P < .02
Median eminence**			
dopamine	239.1 ± 15.9	152.2 ± 11.3	P < .005
norepinephrine	46.9 ± 8.6	58.6 ± 8.5	not sig.
Posterior pituitary**			
dopamine	7.7 ± 0.7	4.4 ± 0.6	P < .01

* determined by fluorescence *(Finch, 1973)*
** not published

(chiefly putamen, *Broch and Marsden 1972)*, median eminence of the hypothalamus, and posterior region of the pituitary (Table 2). We have not found any brain region with age-related reductions of norepinephrine. Reductions of median eminence dopamine and norepinephrine (both) were also found in aging rats *(Miller et al., 1976; Simpkins, Meuller, Huang, and Meites, 1977)*.

The reduced levels of dopamine suggest alteration of two completely distinct dopaminergic projections: the nigro-striatal pathway *(Anden, Carlsson and Dahlstrom, 1964)* and the arcuate-median eminence-hypophyseal system *(Bjorklund, Moore, Nobin and Stenavi, 1973)*. Although there is some evidence for an influence of the substantia nigra on median eminence dopamine *(Kizer, Palkovits and Brownstein, 1976; Palkovits, Fekete, Makara and Herman, 1977)*, there is no known direct nigral dopaminergic connection to the posterior pituitary. Thus, we suggest that at least two independent dopaminergic systems are altered during aging.

A similar trend for dopamine loss in the striatum occurs during normal aging in humans *(Carlsson and Winblad, 1976; Reiderer and Wuketich, 1976)*. About 1% of striatal dopamine is lost per year of adult life. This suggests that impairment of dopaminergic functions is a general characteristic of aging in short- and long-lived mammals. Although data are still fragmentary, various sources indicate that dopaminergic levels and mechanisms are generally more impaired during aging than are functions of other neurotransmitters *(Jonec and Finch, 1975; Finch, 1977b)*. The implications of these and other impairments will be discussed below.

NEUROTRANSMITTER TURNOVER

Two approaches have been used: the rate of processing of radioactive catecholamine precursors and the rate of decrease of levels after pharmacological blockage of synthesis.

PROCESSING OF RADIOACTIVE PRECURSORS

A series of experiments with C57BL/6J male mice aged 12 and 28 mo showed that there was reduced (20 to 50%) conversion of i.p. injected ^{3}H-L-tyrosine or ^{3}H-L-DOPA (dihydroxyphenylalanine) to ^{3}H-catecholamines (chromatographically purified on Alumina and Dowex 50) in the cerebellum, brain stem, hypothalamus and striatum *(Finch, 1973;* Figure 2*)*. Similar impairments in the conversion of ^{3}H-tyrosine (introduced by intra-ventricular cannula) to catecholamines occur in the hypothalamus and striatum of 33 vs 3 mo old rats *(Ponzio, Nicoletta, and Algeri, 1978)*. These widely distributed changes are apparently not a consequence of impaired availability of precursors to the brain because there was no corresponding deficit: (1) of acid soluble ^{3}H in the same brain samples; (2) of the incorporation of ^{3}H-tyrosine into protein in the same brain samples; (3) of the levels of plasma ^{3}H or ^{3}H-L-DOPA injection (Figure 3) *(Finch, 1973)*; or (4) of the uptake of ^{3}H-tyrosine

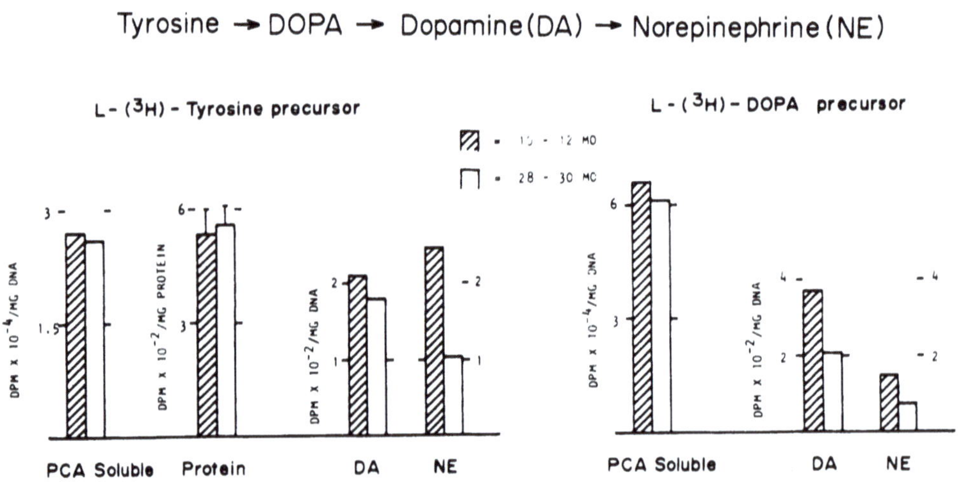

Figure 2. The effect of age on conversion of ^{3}H-tyrosine and ^{3}H-L-DOPA to catecholamines in the hypothalamus of C57BL/6J male mice. Catecholamines were purified by chromatography on Alumina and Dowex (Finch, 1973).

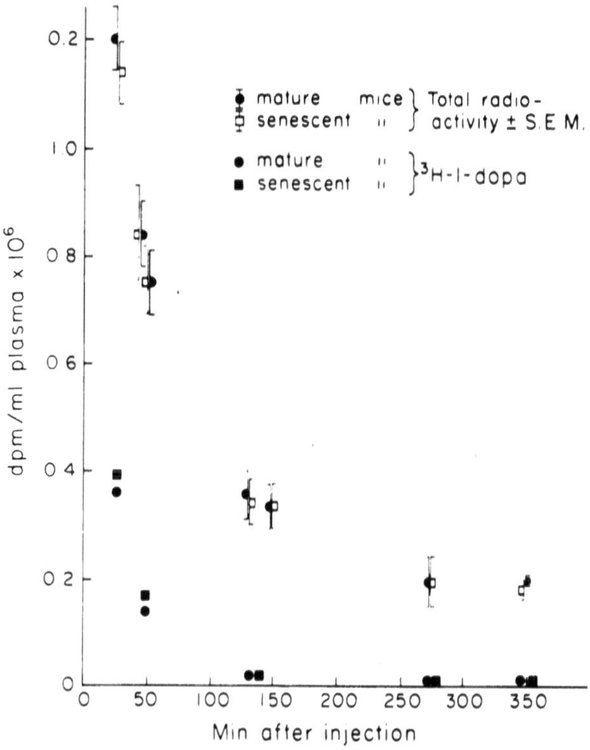

Figure 3. Plasma radioactivity after injection of ^{3}H-L-DOPA (0.75 mg/100 g body weight). ^{3}H-L-DOPA was determined by chromatography on Alumina (Finch, 1973).

or ^{3}H-L-DOPA *in vitro* by slices of the hypothalamus or striatum *(Finch, Jonec, Hody, Walker, Morton-Smith, Alper, Dougher, 1975).*

A study of 9 and 24 mo old male rats (Sprague-Dawley origin) showed small-age-related reductions in basal levels of L-DOPA and 5-HTP (5-hydroxytryptophan) in whole brain (cerebral hemispheres plus cerebellum); after electroconvulsive shock, L-DOPA was elevated to a significantly lesser degree in the 24 mo old rats *(McNamara, Miller, Benignus and Davis, 1977).* There were no age-related differences in 5-HTP, tyrosine, or tryptophan levels. The results of this study, using an entirely different paradigm and a different species, appear to be consistent with our observations of impaired metabolism of ^{3}H-tyrosine and ^{3}H-L-DOPA in aging mice. In view of the region-specific alterations of catecholamines levels during aging, the extensive changes in precursor metabolism observed by us *(Finch, 1973)* and others *(McNamara et. al., 1977; Ponzio et al., 1978)* are surprising. Impairments of energy metabolism in the C57BL/6J mouse brain were found during ischaemia *(Ferendelli,*

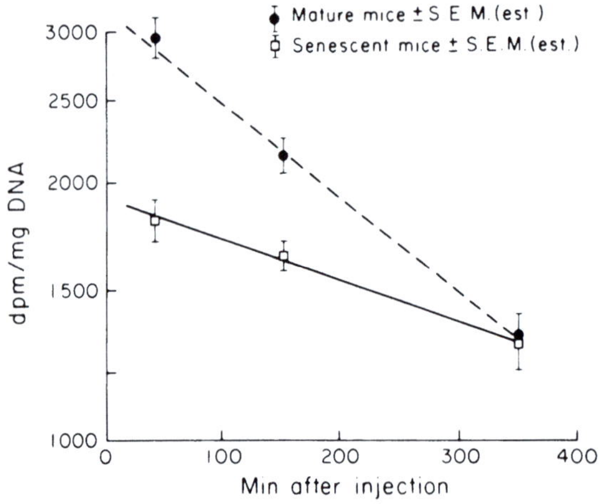

Figure 4. Turnover of ^{3}H-NE in the hypothalamus after injection with ^{3}H-L-DOPA (see Figure 3) and chromatography on Alumina and Dowex. Half-life ± 95% confidence intervals in minutes were: 122 ± 82 (mature), 315 ± 106 (senescent). A second experiment gave a similar age-difference (Finch, 1973).

Sedgwick and Suntzeff, 1971; Maker, Lehrer, Silides and Weiss, 1973). Such changes, whatever their causes, by possibly limiting biosynthetic reserves during stress or physiological demand, could account for smaller increases of L-DOPA in aging rats following electroconvulsive shock *(McNamara et al., 1977).*

Because there were no apparent age differences between 12 and 28 mo old mice in plasma ^{3}H-L-DOPA levels 25 to 350 min after i.p. injection, the turnover of ^{3}H-catecholamines was compared during this time. There was slowed turnover of ^{3}H-norepinephrine in the hypothalamus (Figure 4) based on measurements at three time points (30, 150 and 350 min). Similarly, turnover of ^{3}H-dopamine in the striatum was slowed by 30% *(Finch, 1973).* Compartments with fast turnover would not have been detected. In general, these results are consistent with the reduced conversion of precursors to catecholamines: assuming the steady state, reduced conversion of precursors to catecholamines should be associated with reduced turnover of the labelled catecholamine stores.

BLOCKADE OF SYNTHESIS

The tyrosine analogue, α-methyl-p-tyrosine, causes transient *in vivo* inhibition of catecholamine synthesis by inhibiting tyrosine hydroxylase *(Doteuchi et al., 1974).* This technique, applied to 3 to 4 and 21 mo old male Wistar rats, confirmed the trend for slowed turnover detected by processing of radioactive catecholamine precursors. Based on 1 time point after injection of α-methyl-p-tyrosine, the turnover of hypothalamic dopamine and norepinephrine was slowed by about 30% *(Simpkins, Mueller, Huang and Meites, 1977)*.

SYNAPTOSOMAL UPTAKE

A key control over catecholamine action at the synapse is the reuptake mechanism: it is widely held that there are presynaptic uptake mechanisms in the brain which appear to be analogous to those operating peripherally *(Iverson and Glowinski, 1966)*. Preparations of synaptosomes ("pinched-off nerve endings") show some degree of regional specificity *(Snyder and Coyle, 1969; Shaskan and Snyder, 1970)*. Using crude synaptosomes, we found indications of selective impairments with aging *(Jonec and Finch, 1975)*. In the hypothalamus, dopamine uptake was impaired by age by *ca.* 30%, whereas uptake of norepinephrine, serotonin and tyrosine was unaltered. No age changes in DA leakage from synaptosomes were found which could account for the reduced accumulation of DA observed. Sun *(1976)* also reports that norepinephrine uptake by synaptosomes from the cerebral cortex of C57BL/6J mice is not impaired with age. The selective impairment of dopamine uptake, taken together with the selective reduction of dopamine, could be accounted for by a loss of presynaptic, dopaminergic terminals (discussed below).

Age-related changes in forebrain synaptosomal properties were recently described for 2 *vs* 12 mo old mice: uptake of dopamine, norepinephrine and choline were less, whereas leakage of these substances was greater; no age differences in GABA uptake or leakage were found *(Haycock, White, McGaugh, and Cotman, 1977)*. Because the oldest mice in this study were equivalent to the *youngest* mice in our study, it is difficult to make a general conclusion, beyond the important fact that impairments were selective in both cases.

DOPAMINE ACTIVATED ADENYL CYCLASE

The ability of dopamine to "activate" adenyl cyclase (1.5-2x increases of cAMP levels) in striatal slices or homogenates has received particular attention, because this effect is blocked at low concentration by some dopamine antagonists, e.g. neuroleptics such as butyrophenones or phenothiazines. Aging appears to significantly impair the dopamine-activated adenyl cyclase in the striatum of rats *(Walker and Boas-Walker, 1973; Puri and Volicer, 1976; Govoni, Lodob, Spano and Trabuchi, 1977; Schmidt and Thornberry, 1978)* and rabbits *(Makman, et al., this symp.)*. There was no alteration of basal cAMP in the striatum *(Puri and Volicer, 1976; Schmidt and Thornberry, 1978)*. In contrast, the DA activated cyclase in the retina shows no age-related impairments in rats *(Govoni, et al., 1977)* and rabbits *(Makman et al., this symp.)*. If these findings are physiologically relevant, they could imply impaired effectiveness of the synaptic action of dopamine, e.g. a decoupling or desensitization of dopamine receptors and their interactions with adenyl cyclase.

DOPAMINE RECEPTORS

Recent availability of high specific activity tritiated dopamine agonists (^{3}H-apomorphine) and antagonists (^{3}H-haloperidol, and ^{3}H-spiroperidol) have produced a new research area: characterization of membrane bound dopaminergic receptors. As in the case of the dopamine-activated adenyl cyclase, the relationships of the drug binding sites measured *in vitro* to dopaminergic functions *in vivo* are not yet clear. However, the rankings of drugs for clinically effective doses and their rankings as competitors of binding *in vitro* are intriguingly similar *(Creese, Burt and Snyder, 1976; Seeman, Lee, Chau-Wong and Wong, 1976)*. In the striatum, there is evidence for DA receptor types which are presynaptic *(e.g. Nagy, Lee, Seeman and Fibiger, 1978)* as well as postsynaptic *(Schwarcz, Creese, Coyle and Snyder, 1978)*.

Significant age-related decreases in the levels of specific ^{3}H-spiroperidol binding to membrane preparations from the striatum and hypothalamus have been detected, e.g. *ca.* 40% decreases in the number of specific binding sites (B_{max}) in striatal membrances of 3, 8 and 28 mo old C57BL/6J mice *(Seversen and Finch, in prep.)* and in striatal and hypothalamic preparations of the 5 mo and 60 mo old rabbit *(Makman et al., this symp.)*. No changes in the affinity constant, dissociation rate, or stereospecificity of spiroperidol binding were observed. It is intriguing that the loss of spiroperidol binding is progressive during the adult lifespan. Such continuous changes support the existence of neurochemical aging processes, as distinct from abrupt changes, which might be difficult to dissociate from consequences of age-related diseases.

ORIGINS OF CHANGES IN CATECHOLAMINE REGULATION

DENERVATION OR NEURONAL LOSS

Neuronal loss has long been emphasized to underlie aging phenomena in the brain. With greater recent attention given to the incidence of major pathological lesions (e.g. stroke in the specimens studied), the present evidence suggests that neuronal loss is selective: brain regions or nerve tracts in which neuronal loss is < 15% are illustrated in Table 1. Although data are yet quite limited, trends for age-related loss of neurones have been detected in the substantia nigra *(McGeer, this symp.)* and locus coeruleus *(Brody, 1976)* in humans. If validated by further observations, loss of neurones in the substantia nigra could contribute to striatal dopaminergic aging, whereas loss of neurones in the locus coeruleus could contribute to aging of noradrenergic functions in many parts of the brain *(Swanson and Hartman, 1976)*. We are surveying the extent of neuronal loss during aging in the arcuate nucleus in male C57BL/6J mice: by ordinary histological procedures, there was no loss of neurones (large cresyl violet stained cells) within a margin of 10% between groups of mice aged 12 and 28 mo *(Felicio*

TABLE 3. *COMPARISON BETWEEN STRIATAL CHANGES DURING AGING AND THE ACUTE EFFECTS OF NIGRO-STRIATAL LESIONS IN YOUNG RODENTS*

	AGING	ACUTE EFFECTS OF NIGRO-STRIATAL LESIONS
Dopamine levels	↓ human[1,2] mouse[3]	↓[4]
Tyrosine hydroxylase	↓ human[5], rat[6,7] mouse (?)[7]	↓[8]
Synaptosomal uptake of DA	↓ mouse[9]	↓[10]
Dopamine Receptors (Haloperidol, spiro-peridol binding)	↓ mouse[11], rabbit[12]	↑[13]
Conversion of ^{3}H-tyrosine 2 DA		
Total ^{3}H-DA	↓ mouse[14]	↓[15]
Specific Activity	↓ mouse[14]	↑[15]
Turnover of ^{3}H-DA	↓ mouse[14]	
Adenyl Cyclase (DA-activated)	↓ rabbit[12] ↓ rat[16,17,18,19]	↑[20,21]

The arrows indicate the direction of change.

1 Carlsson and Winblad, 1976
2 Riederer and Wuketich, 1976
3 Finch, 1973
4 Agid et al., 1973
5 McGeer and McGeer, 1976
6 McGeer et al., 1971
7 Reis et al., 1977
8 Agid et al., 1973
9 Jonec and Finch, 1975
10 Kato et al., 1978
11 Severson and Finch, in prep.
12 Makman et al., this symp.
13 Creese et al., 1977
14 Finch, 1973
15 Agid et al., 1973
16 Walker and Boas-Walker, 1973
17 Puri and Volicer, 1976
18 Govoni et al., 1977
19 Schmidt and Thornberry, 1978
20 Von Voigtlander et al., 1973
21 Mishra et al., 1974

and Finch, unpublished observations). At this early stage of analysis, we can only say that there is no evidence for gross neuronal loss in the mouse arcuate during aging: the fate of dopaminergic cells in the arcuate (a minority of the total neuronal population) remains unknown. Given the observed losses of dopamine (about 30%) in the median eminence (see above) it will be of great interest to learn if there are corresponding losses of dopaminergic cell bodies.

It is of interest to compare the various types of age changes reported to the effects of experimental denervation in young rodents. Although most studies employing chemical lesions (6-hydroxydopamine), electrolytic lesions, or surgical lesions causing denervation by neuronal death are short term relative to aging, there are clear distinctions between the effects of aging, and denervation which suggest that quite different mechanism are involved in aging. The best detailed system for comparison between aging and denervation is the nigro-striatal pathway. The available information is summarized in Table 3. The major apparent differences are that (1) dopamine synthesis and turnover is slowed with age (above section), whereas dopamine synthesis is *accelerated* after "denervation" by nigro-striatal lesions *(Agid, Javoy and Glowinski, 1973)*, (2) binding of butyrophenones (^{3}H-haloperidol, ^{3}H-spiroperidol) by striatal membranes (an assay for postsynaptic dopamine receptors) is reduced with age, whereas it is *increased* after "denervation" *(Creese, Burt and Snyder, 1977)* and (3) dopamine-activated adenyl cyclase is impaired with age, whereas it is probably *increased* after denervation *(Von Voigtlander, Boukma and Johnson, 1973; Mishra, Gradner, Katzman and Makman, 1974)*.

Taken together, the available information argues that some phenomena of striatal aging do not conform to the acute effects of nigro-striatal denervation. Several major alternatives may be considered as mechanisms of aging, in addition to nigral neurone loss. (1) There may be a loss of intra-striatal neurones e.g. the kainic acid-induced destruction of striatal interneurones reduces dopamine-activated adenyl cyclase, ^{3}H-haloperidol binding, glutamic acid decarboxylase and choline acetyltransferase activity *(Schwarcz et al., 1978)* in parallel with the aging trend. Moreover, striatal glutamic acid decarboxylase shows age-related decreases in humans of both sexes *(McGeer and McGeer, 1976)* but may not in rats *(McGeer, Fibiger, McGeer and Wickson, 1971)*. (2) Most denervation or lesion studies are acute (1 to 30 weeks), which is short by comparison with the laboratory rodent lifespan. Possibly, a slow denervation process, continuing over a period of 1 to 2 years, has very different effects than does an acute denervation process. (3) Additional caveats include the fact that the evidence in Table 3 is drawn from many sources, not all of which concur (e.g. the evidence for changes in tyrosine hydroxylase); the number of variables in these studies is probably large, extending beyond variations in experimental protocol to matters of animal husbandry, genetic differences, etc.

LOSS OF SYNAPSES

Loss of synapses during aging is observed by the Golgi technique in a number of loci including the cerebral cortex of the rat *(Feldman and Dowd, 1975; Feldman, 1976)* and human *(Scheibel, Lindsey, Tomiyasu and Scheibel, 1975)*, whereas losses were not observed by electron microscopy in human cerebral cortex *(Cragg, 1975)*. Significant losses observed by electron microscopy occur in the dentate gyrus (molecular layer) of aging rats *(Bondareff and Geinisman, 1976; Geinisman and Bondareff, 1976)*. Synaptic loss could account for the decrease of tyrosine hydroxylase, dopamine, and decreased synaptosomal uptake of dopamine in the nigro-striatal projections. Although present ignorance on how synaptic number is regulated in the adult precludes any detailed speculation, it is possible that changes in axonal flow with age (e.g. *Geinisman, Bondareff and Tesler, 1977)* could underlie changes at the synaptic level.

DO GENERALIZED IMPAIRMENTS OF MACROMOLECULAR BIOSYNTHESIS EXIST?

Recent studies argue against the possibility of general age-related changes in brain macromolecular biosynthesis. For example, 1-dimensional gel-electrophoresis studies of the major polypeptides from such brain regions as the hippocampus, hypothalamus, cerebellum, cerebral cortex etc. revealed few quantitative changes and no qualitative ones *(Gordon and Finch, 1974; Vaughan and Calvin, 1977)*. However, such studies detect at best 100 different polypeptides, which is less than 1% of the number present in the brain predicted from messenger RNA complexity *(Bantle and Hahn, 1976)*. Characterization of whole brain RNA by RNA-driven hybridization reactions to nonrepetitive DNA sequence has not revealed any age differences in the yield complexity and sequence representation of nuclear poly-(A)RNA or polysomal poly(A)RNA *(Colman, Osterburn, Kaplan, and Finch, unpublished observations)*, of male Fischer 344 and Sprague-Dawley rats aged 2-32 mo. It appears that the adult pattern of transcription is maintained throughout the rat lifespan in most brain cells. These studies appear to rule out a generalized genomic catastrophy as a major event of aging: if transcription is impaired, the extent of impairment is less than 10% of the total genes transcribed in the young rat brain. Although such a change would still involve an enormous number of different genes (perhaps thousands), the emerging picture is quite different from that suggested by the various error theories of aging *(reviewed in Strehler, 1977)*. It is my view that, if alterations in biosynthesis occur with age in the brain, they will be selective: the decreases of striatal RNA in the mouse *(Chaconas and Finch, 1973)* and rat *(Shaskan, 1977)* contrast with the absence of change in other brain regions.

NEUROENDOCRINE FEEDBACK AND NUTRITIONAL INFLUENCES

Mechanisms distinctly different from cell death or synaptic loss may underlie some neurotransmitter changes. The sensitivity of hypothalamic monoamine metabolism to blood hormone levels is well documented: gonadal *(Anton-Tay, Anton and Wurtman, 1970; Chiocchio, Negro-Vilar and Tramizzani, 1976; Lofstrom, 1977; Luine and McEwen, 1977)*; thyroid *(Engstrom, Svensson and Waldeck, 1974; Jacoby, Mueller and Wurtman, 1975)*. It is thus possible that some monoamine changes represent a response to altered hormone regulation (levels or temporal pattern of secretion). Wilkes *(in this volume)* has found that elevated dopamine in the median eminence of the 24 mo old, constant-estrous, Long-Evans rat is reduced by ovariectomy, thus suggesting that constantly elevated plasma estradiol is involved. In the case of aging changes reported above in male C57BL/6J mice, the absence of detectible changes in the distribution of plasma testosterone values *(Nelson et al., 1975; Finch et al., 1977)* minimizes the possibility that changes of testosterone are the cause of the hypothalamic catecholamine alterations. Many possible endocrine alterations remain to be considered in this context, e.g. thyroid, adrenal cortex etc. Another group of possible factors includes influences of diet on neurotransmitters *(Hutson, Knott and Curzon, 1976; Wurtman and Fernstrom, 1976; Gibson and Wurtman, 1977)*. Even though an older mammal may not show changes in body weight, the reduced caloric intake reported during aging could alter the plasma amino acid profile (levels and diurnal changes) and, in turn, could alter brain monoamine metabolism.

IMPLICATIONS

RELATIONSHIP TO THE LOSS OF OVARIAN CYCLES

The deficits in hypothalamic catecholamine metabolism may be factors in some endocrine alterations of aging. The case is best made for the loss of regular ovarian (estrous or menstrual) cycles. In most laboratory rodents and women, regular ovarian cycles and fertility are lost during midlife *(Finch, 1976 and 1978)*. A variety of agents, including the dopaminergic agonists L-DOPA, lergotrile and iproniazid, have the intriguing ability to reactivate ovarian hormonal (and probably ovulatory) cycles in rats *(Clemens, Amenomori, Jenkins and Meites, 1969; Huang and Meites, 1975; Clemens and Bennett, 1976; Linnoila and Cooper, 1976; reviewed in Finch, 1978)*. Clinical reports of postmenopausal women taking L-DOPA for Parkinson's disease (who showed periodic uterine bleeding) *(Ansel, 1970; Kruse-Larsen and Garde, 1971; Wajsbort, 1972)* suggest that ovarian reactivation may also occur in humans. The dose effective in rats (100 to 200 mg daily s.c.) is equivalent to 10 to 15 gm in human adults; this dose is larger than generally given orally to parkinsonians. These studies led to the hypothesis that the dopaminergic drugs act by compensating for age-related deficits in hypothalamic catecholamines *(Quadri, Kledzik and Meites, 1973; Finch, 1973)*. Although age-related reductions of hypothalamic catecholamine levels and turnover described here are consistent with this hypothesis, other diverse agents including progesterone *(Everett, 1940)*, ACTH and ether stress *(Huang and Meites, 1975)* also can reinitiate cycles. It is at least clear that ovarian oocyte depletion *(reviewed in Talbert, 1977)* is not the only factor in the loss of ovarian cyclicity during aging.

Although the explanation that dopaminergic drugs reactivate ovarian cycles by temporarily compensating for central catecholamine deficiences, it can not be ruled out that other actions are involved: e.g. via the rich sympathetic innervation of the female reproductive tract *(Marshall, 1973; Bahr, Kao and Nalbandov, 1974)*, or via adrenal steroids, such as progesterone. Further exacting experiments are required to show if the age-related deficiency of *hypothalamic* catecholamines is the *cause* of the loss of cycles. A report that peripheral blockers of DOPA decarboxylase (MK-486 or Ro 4-4602) do not prevent L-DOPA from reactivating estrous cycles in aging rats *(Linnoila and Cooper, 1976)* supports the hypothesis of a central action of L-DOPA in reactivating estrous cycles.

It is pertinent that the decreases of hypothalamic catecholamines (20-30%) observed in the aging C57BL/6J male mouse occur without any detectable impairment of gonadal function [plasma testosterone, testicular weight, response of testes to LH *in vivo*, plasma LH, FSH, prolactin and TSH] are all unaltered in 28 mo old C57BL/6J mice *(Finch et al., 1977)*. Thus, reductions of hypothalamic catecholamine levels or turnover are not always correlated with gross deficiencies of hypothalamic and pituitary functioning. Similarly, massive (70 to 80%) deficiencies of striatal dopamine may occur in some Parkinson's subjects without major extra-pyramidal signs *(Bernheimer, Birkmayer, Hornykiewicz, Jellinger and Seitelberger, 1973)*. Nonetheless, deficits of neurotransmitters could impair the mechanism of the preovulatory surge of releasing hormones and gonadotropins without altering the basal (tonic) output of the same hormones.

SLEEP AND BIORHYTHMS

The organization of sleeping-waking activity during the 24 hr (day-night) cycle is altered progressively during normal aging in humans *(Feinburg, 1976)* and rodents *(Richter, 1922; Pittendrigh and Daan, 1974)*. The extensive studies of Feinburg and colleagues show that the number of spontaneous awakenings and time spent awake in bed is positively correlated with age in adults; additionally, there is a 50% decrease during adult life in the total duration of stage 4 or slow wave sleep (the deepest stage with respect to the threshold for arousal). There are also possible relationships between the extent of sleep disturbance in older subjects and cognitive functions *(reviewed in Feinburg, 1976)*.

The possible role of age-related changes in neurotransmitters in the changes of sleep is suggested by many observations in the relationships of monoamine functions to the organization of sleep phases in animal models *(Jouvet, 1969; Putkonen, 1974; Rubin, Poland, Ruben and Gouin, 1974)*. For example, impairment of serotonergic functions by lesions in the raphe *(Jouvet, 1969)* or by inhibition of serotonin synthesis with p-chlorophenylalanine lead to decreased slow wave sleep and insomnia. It would be of obvious interest to know if brain stem serotonin is also reduced in the normal aging human, as in the aging rat *(Meek, Bertilsson, Cheney, Zsilla and Costa, 1977)*.

A role of norepinephrine, dopamine and acetylcholine containing nerves has also been inferred in the regulation of sleep *(Kovacevic and Radulovacki, 1976)*. The possible loss of locus coeruleus neurones (noradrenergic) during human aging *(Brody, 1976)* may be an important factor in sleep changes. The role of neurotransmitters in age-related changes of sleep organization is still highly speculative.

AGING AND NEUROLOGICAL DISEASES

Alzheimer's disease (senile dementia) disease and Parkinson's disease are now identified with major changes in neurotransmitter functions *(reviewed in Barbeau; Carlsson; and Davies, in this volume)*. It is of interest to consider similarities and differences between these diseases and the age-related trends which are emerging. For example, Parkinson's disease is associated with massive deterioration of the nigro-striatal dopamine systems, whereas relatively smaller changes occur during normal aging. The possibility that Parkinson's disease may represent exaggerated or accelerated aging of dopaminergic function has been commented on *(Barbeau, 1962; Barbeau, in press; Finch, 1973)*. However, in addition to major dopaminergic impairments, Parkinson's disease may also be associated with major reduction of striatal glutamic acid decarboxylase *(McGeer and McGeer, 1976)*, whereas striatal glutamic acid decarboxylase may be decreased to a lesser extent during aging in normal humans *(McGeer and McGeer, 1976)*. Data of Schwarcz et al. *(1978)* could implicate alterations in the intrinsic striatal neurones (with glutamic acid decarboxylase) during Parkinson's disease, as well as in normal aging. Because some of the same striatal biochemical changes during aging are found to a greater degree during adult forms of Huntington's disease, it may be speculated that the onset of this disease in the 4th and 5th decade involves an acceleration of striatal aging changes.

The changes in choline acetyl transferase (CAT) found in the cerebral cortex during Alzheimer's disease also occur to a lesser degree during normal human aging *(Davies, 5th vol.)*. The extent of CAT loss during normal aging is apparently similar to the loss of dopamine during normal aging. Possibly, there are common events of aging which underlie functions in the cholinergic, GABA-ergic, and dopaminergic systems, and in the transitions which may occur during Parkinson's, Huntington's and Alzheimer's diseases. A secure picture of comparative striatal changes during aging and age-related neurological disease will require more samples by one or more orders of magnitude: various subgroups may become apparent with concurrent measurements of a battery of morphological and neurochemical parameters.

CONCLUSION

The evidence that impairments of monoaminergic mechanisms occur as a part of the normal aging process is derived from more than 10 laboratories. It is now plausible to consider that changes such as those in striatal dopaminergic function are a genetically programed feature of aging in short- and long-lived mammals. The major issues ahead involve the extent of localization of these changes in particular neural systems, the role of neuronal loss in initiating or driving these changes of aging, and the relationships of changes in neurotransmitter regulation to the physiological and pathologicalchanges of age-related disease, including Parkinson's, Huntington's and Alzheimer's disease.

REFERENCES

Agid, Y., Javoy, F. and Glowinski, J. (1973). Hyperactivity of remaining dopaminergic neurones after partial destruction of the nigro-striatal dopaminergic system in the rat. *Nature New Biol. 245,* 150-151.

Anden, N.E., Carlsson, A., Dahlstrom, A., Fuxe, F., Hillarp, N.A., and Larsson, K. (1974). Demonstration and mapping out of nigro-neostriatal dopamine neurones. *Life Sci., 3,* 523-530.

Ansel, R.D. (1970). In: *L-DOPA and Parkinsonism.* pp. 317. Eds. Barbeau, A. and McDowell, F.H., Philadelphia.

Anton-Tay, F., Anton, S.M. and Wurtman, R.J. (1970). Mechanism of changes in brain norepinephrine metabolism after ovariectomy. *Neuroendocrinology 6,* 265-273.

Bahr, J., Kao, L. and Nalbandov, A.M. (1974). The role of catecholamines and nerves in ovulation. *Biol. Reprod. 10,* 273-290.

Bantle, J.A. and Hahn, W.E. (1976). Complexity and characterization of polyadenylated RNA in the mouse brain. *Cell 8,* 139-150.

Barbeau, A. (1962). The pathogenesis of Parkinson's disease: A new hypothesis. *Can. Med. Assoc. J. 87,* 802-807.

Barbeau, A. (1978). Peptides in Parkinson's disease. In: *Aging and Neuroendocrine Relationships*, C.E. Finch, D.E. Potter and A.D. Kenny, Plenum Press, New York.

Bernheimer, H., Birkmayer, W., Hornykiewicz, O., Jellinger, K. and Seitelberger, F. (1973). Brain dopamine and the syndromes of Parkinson and Huntington. Clinical, morphological and neurochemical correlations. *J. Neurol. Sci. 20,* 415-455.

Björklund, A., Moore, R.Y., Nobin, A. and Stenevi, U. (1973). The organization of tubero-hypophyseal and reticulo-infundibular catecholamine neuron systems in the rat brain. *Brain Res. 51,* 171-191.

Bondareff, W. and Geinisman, Y. (1976). Loss of synapses in the dentate gyrus of the senescent rat. *Am. J. Anat. 145,* 129-136.

Broch, O.J. and Marsden, C.A. (1972). Regional distribution of monoamines in the corpus striatum of the rat. *Brain Res. 38,* 524-428.

Brody, H. (1970). Structural changes in the aging nervous system. In: Interdisciplinary Topics in Gerontology 7, 9-21. Karger, Basal.

Brody, H. (1976). An examination of cerebral cortex and brain stem aging. In: *Neurobiology of Aging*, pp. 177-182. Eds, Terry, R.D. and Gershon, S. Raven Press, New York.

Bronson, F.H. and Desjardins, C. (1977). Reproductive failure in aged CBF_1 male mice: Interrelationships between pituitary gonadotropic hormones, testicular function and mating success. *Endocrinology, 101*, 939-945.

Carlsson, A. and Winblad, B. (1976). The influence of age and time interval between death and autopsy on dopamine and 3-methoxy tyramine levels in human basal ganglia. *J. Neurol. Transmiss. 38*, 271-276.

Carlsson, A. (1978). Age-dependent changes in central dopaminergic and other monoaminergic systems. In: *Aging and Neuroendocrine Relationships*, Eds. Finch, C.E., Potter, D.E. and Kenny, A.D., Plenum Press, New York

Chaconas, G. and Finch, C.E. (1973). The effect of aging on RNA/DNA ratios in brain regions of the C57BL/6J male mouse. *J. Neurochem. 21*, 1469-1473.

Chiocchio, S.R., Negro-Vilar, A. and Tramezzani, J.H. (1976). Acute changes in norepinephrine content in the median eminence induced by orchidectomy or testosterone replacement. *Endocrinology 99*, 629-635.

Clemens, J.A., Amenomore, Y., Jenkins, T. and Meites, J. (1969). Effects of hypothalamic stimulation, hormones and drugs on ovarian function in old female rats. *Proc. Soc. exp. Biol. Med. 132*, 561-563.

Coleman, G.L., Barthold, S.W., Osbaldiston, G.W., Foster, S.J. and Jonas, A.M. (1977). Pathological changes during aging in barrier-reared Fischer 344 male rats. *J. Geront. 32*, 258-278.

Corbin, K.B. and Gardner, E.D. (1937). Decrease in number of myelinated fibres in human spinal root nerves with age. *Anat. Rec. 68*, 63-74.

Coyle, J.T. and Henry, D. (1973). Catecholamines in fetal and newborn rat brain. *J. Neurochem. 21*, 61-67.

Cragg, B.G. (1975). The density of synapses and neurons in normal, mentally defective and aging human brains. *Brain 98*, 81-90.

Creese, I., Burt, D.R. and Snyder, S.H. (1976). Dopamine receptor binding predicts clinical and pharmacological potencies of antischizophrenic drugs. *Science 192*, 481-483.

Creese, I., Burt, D.R. and Snyder, S.H. (1977). Dopamine receptor binding enhancement accompanies lesion-induced behaviorial sensitivity. *Science 197*, 596-598.

Davies, P. (1978). Loss of Choline Acetyltransferase Activity in Normal Aging and in Senile Dementia. In: *Aging and Neuroendocrine Relationships*. Eds. Finch, C.E., Potter, D.E., Kenny, A.D., Plenum Press, New York.

Doteuchi, M., Wang, C. and Costa, E. (1974). Compartmentation of dopamine in rat striatum. *Mol. Pharmac. 10*, 225-234.

Duchen, L.W. and Schurr, P.H. (1976). The pathology of the pituitary gland in old age. In: *Hypothalamus, Pituitary and Ag-*

ing. pp. 137-156. Eds. Everitt, A.B. and Burgess, J.A. Charles C. Thomas, Springfield, Illinois.

Dunn, T.B. (1954). Normal and pathological anatomy of the reticular tissue in laboratory mice, with a classification and discussion of neoplasm. *J. Natn. Cancer Inst. 14,* 1281-1433.

Durbin, P.W., Williams, M.H., Jeung, N. and Arnold, J.S. (1966). Development of spontaneous mammary tumors over the life-span of the female Charles River (Sprague-Dawley) rat: the influence of ovariectomy, thyroidectomy, and adrenalectomy-ovariectomy. *Cancer Res. 26,* 400-411.

Engström, G., Svensson, T.H. and Waldeck, B. (1974). Thyroxine and brain catecholamines: increased transmitter synthesis and increased receptor sensitivity. *Brain Res. 77,* 571-483.

Everett, J.W. (1940). The restoration of ovulatory cycles and corpus luteum formation in persistent-estrous rats by progesterone. *Endocrinology 27,* 681-686.

Feinberg, I. (1976). Functional implications of changes in sleep physiology with age. In: *Neurobiology of Aging.* pp. 23-41. Eds. Terry, R.D. and Gershon, S., Raven Press, New York.

Feldman, M.L. and Dowd, C. (1975). Loss of dendritic spines in aging cerebral cortex. *Anat. Rec. 148,* 279-301.

Feldman, M.L. (1976). Aging changes in the morphology of cortical dendrites. In: *Neurobiology of Aging.* pp. 211-227. Eds. Terry, R.D. and Gershon, D., Raven Press, New York.

Ferrendelli, J.A., Sedgwick, W.G. and Suntzeff, V. (1971). Regional energy metabolism in lipofuscin in accumulation in mouse brain during aging. *Neuropathology 30,* 638-649.

Finch, C.E. (1969). Cellular activities during aging in mammals. Ph.D. Thesis. The Rockefeller University, New York.

Finch, C.E., Foster, J.R. and Mirsky, A.E. (1969). Aging and the regulation of cell activities during exposure to cold. *J. Gen. Physiol. 54,* 690-712.

Finch, C.E. (1971). The comparative biology of senescence: some evolutionary and developmental considerations. In: *Animal Models for Biomedical Research.* pp. 47-67. National Academy of Sciences (US).

Finch, C.E. (1972). Enzyme activities, gene function and aging in mammals. *Exp. Geront. 7,* 53-67.

Finch, C.E. (1973). Catecholamine metabolism in the brains of aging male mice. *Brain Res. 52,* 261-276.

Finch, C.E. and Foster, J.R. (1973). Hematologic and serum electrolyte values of the C57BL/6J male mouse in maturity and senescence. *Lab. Animal Sci. 23,* 339-349.

Finch, C.E. and Girgis, F.G. (1974). Enlarged seminal vesicles of senescent C57BL/6J mice. *J. Geront. 29,* 134-138.

Finch, C.E., Jonec, V., Hody, G., Walker, J.P., Morton-Smith, W., Alper, A. and Dougher, G.J. (1975). Aging and the passage of L-tyrosine, L-DOPA and inulin into mouse brain slices in vitro. *J. Geront. 30,* 33-40.

Finch, C.E. (1976). The regulation of physiological changes during

mammalian aging. *Q. Rev. Biol. 51,* 49-83.
Finch, C.E. (1977a). Some considerations regarding disease in old age. *J. Geront. 32,* 642.
Finch, C.E. (1977b). Neuroendocrine and autonomic aspects of aging. In: *Handbook of the Biology of Aging.* Eds. Finch, C.E. and Hayflick, L. Van Nostrand Reinhold, New York, pp. 262-280.
Finch, C.E. and Flurkey, K. (1977). The molecular biology of estrogen replacement therapy. *Contemp. Obstet. Gyn. 9,* 97-107.
Finch, C.E., Jonec, V., Wisner, J.R. Jr., Sinha, Y.N., de Vellis, J.S. and Swerdloff, R.S. (1977). Hormone production by the pituitary and testes of male C57BL/6J mice during aging. *Endocrinology 101,* 1310-1317.
Finch, C.E., (1978). Reproductive senescence in rodents: factors in the decline of fertility and loss of regular estrous cycles. In: *The Aging Reproductive System, Aging vol. 4.* Ed. Schneider, E.L. Raven Press, New York, pp. 193-212.
Geinisman, Y. and Bondareff, W. (1976). Decrease in the number of synapses in senescent brain: A quantitative electron microscopic analysis of the dentate gyrus molecular layer in the rat. *Mech. Aging Dev. 5,* 363-378.
Geinisman, Y., Bondareff, W. and Telse, A. (1977). Diminished axonal transport of glycoproteins in the senescent rat brain. *Mech. Aging Dev. 6,* 363-378.
Gibson, G.J. and Wurtman, R.J. (1977). Physiological control of brain catechol synthesis by brain tyrosine concentration. *Biochem. Pharmacol. 26,* 1137-1142.
Gordon, S.M. and Finch, C.E. (1974). An electrophoretic study of protein synthesis in brain regions of senescent male mice. *Exp. Geront. 9,* 269-273.
Govoni, S., Loddo, P., Spano, P.F. and Trabuchi, M. (1977). Dopamine receptor sensitivity in brain and retina of rats during aging. *Brain Res. 138,* 565-570.
Greisbach, W.E. (1967). Basophil adenomata in the pituitary glands of 2-year old male long-Evans rats. *Cancer Res. 27,* 1813-1818.
Hanna, M.G., Jr., Nettesheim, P. and Snodgrass, M.H. (1971). Decreasing immune competence and development of reticulum cell sarcomas in lymphatic tissue of aged mice. *J. Natn. Cancer Inst. 46,* 809-824.
Haycock, J.W., White, W.F., McGaugh, J.L., and Cotman, C.W. (1977). Enhanced stimulus-secretion coupling from brains of aged mice. *Exp. Neurol. 57,* 873-882
Hökfelt, T. and Fuxe, K. (1972). Effects of prolactin and ergot alkaloids on the tubero-infundibular dopamine (DA) neurones. *Neuroendocrinology 9,* 100-122.
Huang, H.H. and Meites, J. (1975). Reproductive capacity of aging female rats. *Neuroendocrinology 17,* 289-295.
Hutson, P.H., Knott, P.J. and Curzon, G. (1976). Control of brain tryptophan concentration in rats on a high fat diet. *Nature 262,* 142-143.
Iverson, L.L. and Glowinski, J. (1966). Regional studies of cate-

cholamines in the rat brain-II. *J. Neurochem. 13*, 671-682.
Jacoby, J.H., Mueller, G. and Wurtman, J.R. (1975). Thyroid state and brain monoamine metabolism. *Endocrinology 97*, 1332-1335.
Jonec, V. and Finch, C.E. (1975). Aging and dopamine uptake by subcellular fractions of the C57BL/6J male mouse brain. *Brain Res. 91*, 197-215.
Jouvet, M. (1969). Biogenic amines and the states of sleep. *Science 163*, 32-41.
Kato, G., Carson, S., Kemme, M.L., Glowinski, J., and Giorguieff, M.F. (1978). Changes in striatal specific ^{3}H-atropine binding after unilateral 6-hydroxydopamine lesions of nigrostriatal dopaminergic neurons. *Life Sci. 22*, 1607-1614.
Kizer, J.S., Palkovits, M. and Brownstein, M.J. (1976). The projections of the A8, A9, and A10 dopaminergic cell bodies: evidence for a nigral-hypothalamic-median eminence dopaminergic pathway. *Brain Res. 108*, 363-370.
Konigsmark, B.V. and Murphy, E.A. (1970). Neuronal populations in the human brain. *Nature 299*, 1335-1336.
Kovacevic, R. and Radulovacki, M. (1976). Monoamine changes in the brain of cats during slow-wave sleep. *Science 193*, 1025-1027.
Kruse-Larson, C. and Garde, K. (1971). Postmenopausal bleeding: another side effect of levodopa. *Lancet i*, 707-709.
Kunstyr, I. and Leuenberger, H.G.W. (1975). Gerontological data of C57BL/6J mice. I. Sex differences in survival curves. *J. Geront. 30*, 157-162.
Linnoila, M. and Cooper, R.L. (1976). Reinstatement of vaginal cycles in aged female rats. *J. Pharmac. exp. Ther. 199*, 477-482.
Löfstrom, A. (1977). Catecholamine turnover alterations in discrete areas of the median eminence of the 4- and 5-day cyclic rat. *Brain Res. 120*, 113-131.
Luine, V.N. and McEwen, B.S. (1977). Effect of oestradiol on turnover of type A monoamine oxidase in brain. *J. Neurochem. 28*, 1221-1227.
Maker, H.S., Lehrer, G.M., Silides, D.J. and Weiss, C. (1973). Regional energy metabolism during maturation and aging of mouse cerebellum. *Prog. Brain Res. 40*, 293-307.
Makman, M.H., Ahn, H.S., Thal, L.J., Dvorkin, B., Horowitz, S.G., Sharpless, N.S. and Rosenfeld, M. (1978). Biogenic amine stimulated adenylate cyclase and spiroperidol-binding sites in rabbit brain: evidence for selective loss of receptors with aging. In: *Aging and Neuroendocrine Relationships*. Eds. Finch, C.E., Potter, D.E. and Kenny, A.D., Plenum Press, New York.
Marshall, J.M. (1973). Effects of catecholamines on the smooth muscle of the female reproductive tract. *Ann. Rev. Pharmac. 13*, 19-32.
McGeer, E.G., Fibiger, H.C., McGeer, P.L. and Wickson, V. (1971). Aging and brain enzymes. *Exp. Geront. 6*, 391-396.

McGeer, P.L. and McGeer, E.G. (1976). Enzymes associated with the metabolism of catecholamines, acetylcholine, and GABA in human controls and patients with Parkinson's disease and Huntington's chorea. *J. Neurochem.* *26*, 65-76.

McGeer, P.L., McGeer, E.G. and Suzuki, J.S. (1977). Aging and extrapyramidal function. *Arch. Neurol.* *34*, 33-35.

McGeer, P.L. and McGeer, E.G. (1978). Aging and neurotransmitter systems. In: *Aging and Neuroendocrine Relationships*. Eds. Finch, C.E., Potter, D.E. , Kenny, A.D., Plenum Press, New York.

McNamara, M.C., Miller, A.T., jr., Benignus, V.A. and Davis, J.N. (1977). Age-related changes in the effect of electroconvulsive shock (ECS) on the in vivo hydroxylation of tyrosine and tryptophan in rat brain. *Brain Res.* *131*, 313-320.

Meek, J.L., Bertillsson, L., Cheney, D.L., Zsilla, G. and Costa, E. (1977). Aging-induced changes in acetylcholine and serotonin content of discrete brain nuclei. *J. Geront.* *32*, 129-131.

Miller, A.E., Shaar, C.J., and Riegle, G.D. (1976). Aging effects on hypothalamic dopamine and norepinephrine content in the male rat. *Exp. Aging Res.* *2*, 475-480.

Mishra, R.K., Gardner, E.L., Katzman, R. and Makman, M.H. (1974). Enhancement of dopamine-stimulated adenyl cyclase activity in rat caudate after lesions in substantia nigra: evidence for denervation supersensitivity. *Proc. Natn. Acad. Sci.* *71*, 3883-3887.

Monagle, R.D. and Brody, H. (1974). The effects of age upon the main nucleus of the inferior olive in the human. *J. Comp. Neurol.* *155*, 61-66.

Moyer, E.K. and Kalizewski, B.F. (1958). The number of nerve fibers in motor spinal nerve roots of young, mature and aged cats. *Anat. Rec.* *131*, 681-699.

Nagy, J.I., Lee, T., Seeman, P., and Fibiger, H.C. (1978). Direct evidence for presynaptic and postsynaptic dopamine receptors in brain. *Nature* *274*, 278-281.

Nelson, J.F., Latham, K. and Finch, C.E. (1975). Plasma testosterone levels in C57BL/6J male mice: effects of age and disease. *Acta Endocr. (Kbh)* *80*, 744-752.

Palkovits, M., Brownstein, M., Saavedra, J. and Axelrod, J. (1974). Norepinephrine and dopamine content of hypothalamic nuclei of the rat. *Brain Res.* *77*, 137-149.

Palkovits, M., Fekete, M., Makara, G.B. and Herman, J.P. (1977). Total and partial hypothalamic deafferentations for topagraphical identification of catecholaminergic innvervations of certain preoptic and hypothalamic nuclei. *Brain Res.* *127*, 127-236.

Pittendrigh, C.S. and Daan, S. (1974). Circadian oscillations in rodents: a systematic increase of their frequency with age. *Science* *186*, 548-550.

Ponzio, F., Brunell, N., and Algeri, S. (1978). Catecholamine synthesis in the brain of aging rats. *J. Neurochem. 30,* 1617-1620.

Puri, S.K. and Volicer, L. (1976). Effect of aging on cyclic AMP levels and adenylate cyclase and phosphodiesterase activities in the rat corpus striatum. *Mech. Aging Dev. 6,* 53-58.

Putkonen, D.S. (1974). Monoamines and the regulation of vigilance and sleep. *Med. Biol. 52,* 193-199.

Quadri, S.K., Kledzik, G.S. and Meites, J. (1973). Reinitiation of estrous cycles in old constant-estrous rats by central-acting drugs. *Neuroendocrinology 11,* 248-255.

Reis, D.J., Ross, R.A. and Joh, T.H. (1977). Changes in the activity and amounts of enzymes synthesizing catecholamines and acetylcholine in brain, adrenal medulla, and sympathetic ganglia of aged rat and mouse. *Brain Res. 136,* 465-474.

Reisine, T.D., Fields, J.Z. and Yamamura, H.I. (1977). Neurotransmitter receptor alterations in Parkinson's disease. *Life Sci. 21,* 335-344.

Richter, C.P. (1922). A behavioristic study of the activity of the rat. *Comp. Psychol. Monographs 1,* 1-55.

Riederer, P. and Wuketich, St. (1976). Time course of nigrostriatal degeneration in Parkinson's disease. *J. Neurol. Transmiss. 38,* 277-301.

Riegle, G.D. and Miller, A.E. (1978). Aging effects on the hypothalamic-hypophyseal-gonadal control system in the rat. In: *The Aging Reproductive System.* pp. 159-192. Ed. Schneider, E.L., Raven Press, New York.

Rubin, R.T., Poland, R.E., Rubin, L.E. and Gouin, P.R. (1974). The neuroendocrinology of human sleep. *Life Sci. 14,* 1041-1052.

Sant'ambrogio, G., Frazier, D. and Boyarsky, L.L. (1961). Alterations in the central latency, motoneurone number and blood volume of spinal cord of the aging rat. *Am. J. Physiol. 200,* 927-930.

Scheibel, M., Lindsay, R.D., Tomiyasu, U. and Scheibel, A.B. (1975). Progressive dendritic changes in human cortex. *Exp. Neurol. 47,* 392-403.

Schmidt, M.J. and Thornberry, J.F. (1978). Cyclic AMP and cyclic GMP accumulation in vitro in brain regions of young, old and aged rats. *Brain Res. 139,* 159-177.

Schwarcz, R., Creese, I., Coyle, J.T. and Snyder, S.H. (1978). Dopamine receptors localized on cerebral cortical afferents to rat corpus striatum. *Nature 271,* 766-768.

Seeman, P., Lee, T., Chau-Wong, M. and Wong, K. (1976). Antipsychotic drug doses and neuroleptic/dopamine receptors. *Nature 261,* 717-718.

Shaskan, E.G. and Snyder, S.H. (1970). Kinetics of serotonin accumulation into slices from rat brain: relationship to catecholamine uptake. *J. Pharmac. exp. Ther. 175,* 404-418.

Shaskan, E.G. (1977). Brain regional spermidine and spermine levels in relation to RNA and DNA in aging rat brain. *J. Neurochem.* *28*, 509-516.

Simpkins, J.W., Mueller, G.P., Huang, H.H. and Meites, J. (1977). Evidence for depressed catecholamine and enhanced serotonin metabolism in aging male rats: possible relation to gonadotropin secretion. *Endocrinology 100*, 1672-1678.

Sinha, Y.N., Selby, R.W. and Vanderlaan, W.P. (1974). The natural history of prolactin and GH secretion in mice with high and low incidence of mammary tumors. *Endocrinology 94*, 757-764.

Snyder, S.H. and Coyle, J.T. (1969). Regional differences in ^{3}H-norepinephrine and ^{3}H-dopamine uptake into rat brain homogenates. *J. Pharmac. exp. Ther. 165*, 78-86.

Staats, J. (1976). Standardized nomenclature for inbred strains of mice: Sixth listing. *Cancer Res. 36*, 4333-4377.

Strehler, B.L. (1977). In: *Time, Cells and Aging*, 2nd Ed. Academic Press, New York.

Sun, A.Y. (1976). Aging and in vivo norepinephrine-uptake in mammalian brain. *Exp. Aging Res. 2*, 207-219.

Swanson, L.W. and Hartman, B.K. (1976). The central adrenergic system: an immunofluorescence study of the location of cell bodies and their efferent connections in the rat utilizing dopamine-β-hydroxylase as a marker. *Comp. Neurol. 163*, 467-506.

Talbert, G. (1977). Aging of the female reproductive system. In: *Handbook of the Biology of Aging*. pp. 318-356. Eds. Finch, C.E. and Hayflick, L., Van Nostrand, New York.

Vaughan, W.J. and Calvin, M. (1977). Electrophoretic analysis of brain proteins from young adult and aged mice. *Gerontologia 23*, 110-126.

Vermeulen, A., Reubens, R. and Verdonck, L. (1972). Testosterone secretion and metabolism in male senescence. *J. Clin. Endocr. Metab. 34*, 730-735.

Versteeg, D.H.G. and Wurtman, R.J. (1975). Effect of $ACTH_{4-10}$ on the rate of synthesis of [^{3}H] catecholamines in the brains of intact, hypophysectomized and adrenalectomized rats. *Brain Res. 93*, 552-557.

Vijayashankar, N. and Brody, H. (1971). The neuronal population of human abducens nucleus. *Anat. Rec. 169*, 447.

Vijayashankar, N. and Brody, H. (1972). The neuronal population of the nuclei of the trochlear nerve and the locus coeruleus in the human. *Anat. Rec. 172*, 421-422.

Von Voigtlander, P.F., Boukma, S.J. and Johnson, G.A. (1973). Dopaminergic denervation supersensitivity and dopamine stimulated adenyl cyclase activity. *Neuropharmacology 12*, 1081-1086.

Wajsbort, J. (1972). Post-menopausal bleeding after L-DOPA. *New Engl. J. Med. 286*, 784.

Walker, J.P. and Boas-Walker, J. (1973). Properties of adenyl cyclase from senescent rat brain. *Brain Res. 54*, 391-396.

Weiner, N. (1974). A critical assessment of methods for the determination of monoamine synthesis turnover rates in vivo. In: *Neuropsychopharmacology of Monoamines and Their Regulatory Enzymes*. Ed. Usdin, E. Raven Press, New York. *Adv. Biochem. Psychopharmac 12*, 143-159.

Wilkes, M.M., Lu, K.H., Fulton, S.L. and Yen, S.S.C. (1978). Hypothalamic-pituitary-ovarian interactions during reproductive senescence in the rat. In: *Aging and Neuroendocrine Relationships*. Eds. Finch, C.E., Potter, D.E. and Kenny, A.D. Plenum Press, New York.

Wright, E.A. and Spink, J.M. (1959). A study of the loss of nerve cells in the central nervous system in relation to age. *Gerontologia 3*, 277-287.

Wurtman, R.J. and Fernstrom, J.D. (1976). Control of brain neurotransmitter synthesis by precursor availability and nutritional state. *Biochem. Pharmac. 25*, 1691-1696.

AGING AND NEUROTRANSMITTER SYSTEMS

P.L. McGEER and E.G. McGEER

Kinsmen Laboratory of Neurological Research
Department of Psychiatry
University of British Columbia
Vancouver, Canada, V6T 1W5

INTRODUCTION

The aging process is accompanied by substantial alterations in physiological function. Decreased total motor activity, declining mental acuity and altered endocrine performance are among the obvious changes that take place. Within the brain, lipofuscin deposits accumulate and the total weight tends to shrink.

It might be anticipated that these gross cellular changes, as well as the obvious changes in neurological function, might be correlated with biochemical measures. Neurotransmitter agents, or the enzymes concerned with their synthesis, are potential indices of such change because they are "markers" for specific neurons and because they reflect something of the state of interneuronal communication. For example, the specific cell losses in the substantia nigra in Parkinson's disease can be correlated with drops in dopamine and its synthetic enzymes tyrosine hydroxylase (TH) and dopa decarboxylase (DDC). Similarly, neuronal losses in the basal ganglia in Huntington's chorea can be correlated with losses in cholinacetyltransferase (CAT); GABA, and its synthetic enzyme glutamic acid decarboxylase (GAD); and Substance P.

If neuronal losses of either a specific or general nature occur during the aging process, then it would be reasonable to anticipate that such losses could be detected by appropriate measurement of neurotransmitter amines and/or their synthetic enzymes. Similarly, any concentration or enhancement of these materials as an adaptation to the aging process should also be detectable by these techniques. In this chapter we will review some of the literature, with particular reference to our own study of brain enzyme acitivities in a series of 28 humans dying from accidents or in hospital from non-neurological illness. We also describe comparable results obtained with Wistar rats up to the elderly rat age of 30 months. The synthetic enzymes rather than the amines themselves were chosen for study because they can be measured reliably on small tissue samples and because they seem somewhat less subject to postmortem changes.

STUDIES IN RATS

A. THE EFFECT OF POSTMORTEM DELAY

The time from death to autopsy varies considerably in human cases. It is an uncontrollable factor which would conceivably

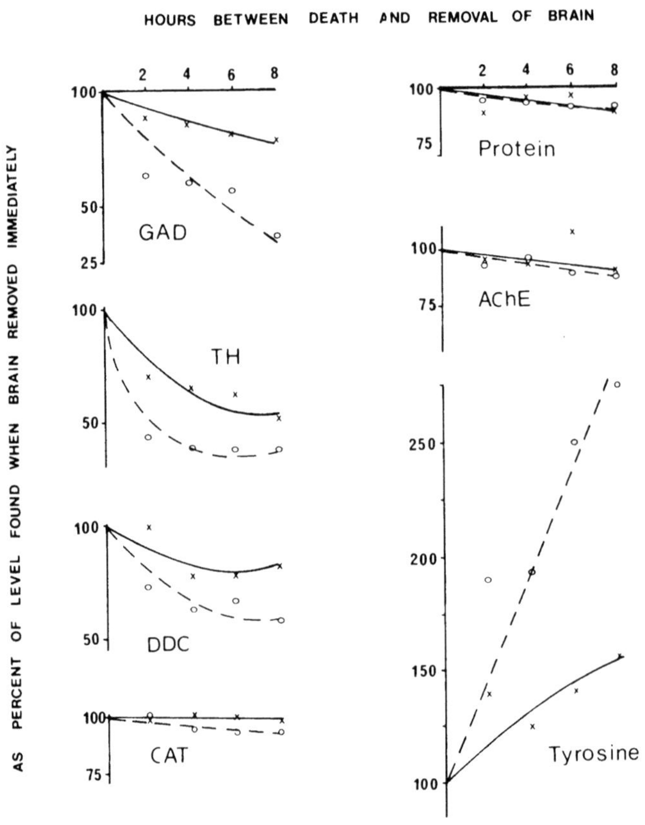

Figure 1. Postmortem changes in some enzyme activities, protein and tyrosine concentrations in rat brain. Rats were sacrificed and left at either 4°C (x-)or 20°C (o---) for indicated number of hours before removal of the brain.

mask any premortem changes associated with age or other factors. As one approach to judging the importance of postmortem delay, we have carried out studies on rats, waiting for different periods of time between sacrifice and enzyme measurement. Figure 1 shows the decrement in synthetic enzyme levels with postmortem delay for GABA, acetylcholine and catecholamine neuronal systems. Also shown are the decrements in acetylcholinesterase (AChE) and total protein, and the increase in tyrosine. Inspection of the figure shows the anticipated more rapid decline in enzymes when the animals are left at room temperature (dotted lines) as opposed to 4°C (solid lines). There is a fairly rapid decline in TH, DDC and GAD. CAT declines much more slowly and, at 4°C the decrement is scarcely noticeable over a period of 8 hours. There is a slow decline in protein probably due to hydrolysis which could also account for the increase in level of the free amino acid tyrosine.

These data indicate that postmortem delay can certainly not be ignored in human studies, but that the effects are not so extreme as to vitiate a carefully executed study, particularly if some record is kept of the speed with which the autopsy is performed in a given case. Experience in several laboratories has indicated, in fact, that postmortem delays of 2-48 hours have little significant effect upon the results *(Bird and Iverson, 1974; Grote, Moses, Robins, Hudgens and Croninger, 1974; McGeer and McGeer, 1976b).*

A more obvious contributing factor to scatter in normal human data appears to be the immediate premortem condition of the patient. Studies performed in our laboratory *(McGeer and McGeer, 1976b)* suggest that a state of coma prior to death, particularly if the patient has been supported by an artificial respirator, can lead to very major declines in many enzyme levels, thereby rendering the data almost valueless.

B. CHANGES IN ENZYME LEVELS IN NEONATAL RATS

Figure 2 shows the postnatal development of rat striatal TH and CAT. As can be seen from the figure, TH at 5 days of age is less than 30% of the adult level. It rises progressively, achieving adult levels by approximately 20 days. The TH in the striatum is, of course, largely contained in dopaminergic nerve endings. Cell bodies lie in the substantia nigra. Brainstem TH does not show a similar pattern of postnatal increase. Instead, it remains relatively constant, no doubt reflecting the fact that the cell bodies are not dividing in the postnatal period *(McGeer, Gibson, Wada and McGeer, 1967; McGeer, Parkinson and McGeer, 1976).* The increase in the striatum, on the other hand, reflects the proliferation of nerve endings in the 5-20 day period *(McGeer, Fibiger and Wickson, 1971).*

TH increases in the striatum precede those in CAT, where there is a sharp increase between the 10th and 20th day postmortem. This is the time of greatest synaptogenesis in the striatum *(Hattori*

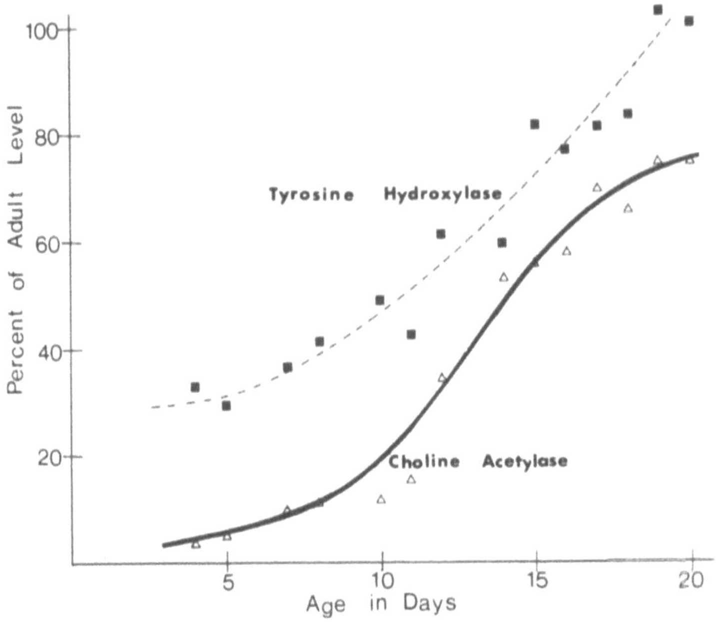

Figure 2. Choline acetyltransferase (choline acetylase, CAT) and tyrosine hydroxylase activities in neonatal rat caudate/putamen as a function of age.

and McGeer, 1973). It has been shown that dopaminergic nerve endings make contact with cholinergic dendritic spines in the striatum *(Hattori, Singh, McGeer and McGeer, 1976).* The developmental data suggest that dopaminergic nerve endings are forming prior to the elaboration of the cholinergic dendritic spines which receive them.

Similar neonatal changes in striatal TH and CAT have been observed by Coyle and Campochiaro *(1976)* in the rat. Parallel changes in dopamine in the neonatal period in the striatum have also been found in the rat *(Lindblatt and Carlsson, this volume)* and in the rabbit *(Tennyson, Barrett, Cohen, Cote, Heikkila and Mytilineou, 1973).* In the latter case, the increases in dopamine levels and fluorescence have been shown to parallel synaptogenesis as seen by electron microscopy.

GAD activity in the striatum shows a slightly less marked increase than TH or CAT in the neonatal period and an almost identical pattern was found in the interpeduncular nucleus - ventral tegmental area (Figure 3). This contrasts with the diverse developmental patterns of TH which shows, as previously mentioned, a much more marked increase in terminal areas, such as the striatum,

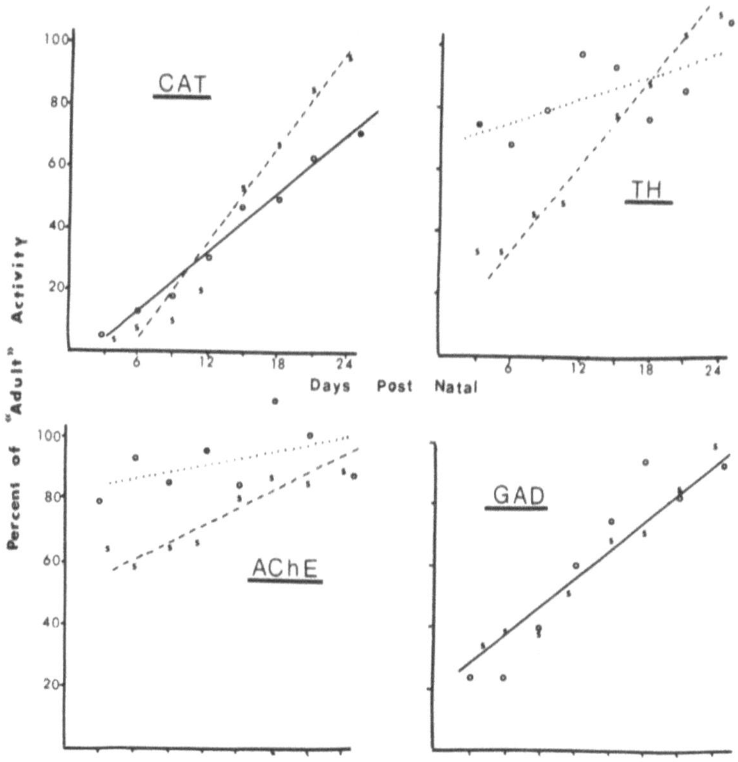

Figure 3. *Mean enzyme activity as percentage of 30-day levels in rat neostriatum (s) and interpeduncular nucleus-ventral tegmentum area (o) plotted as a function of age, with calculated lines of linear correlation for striatum (---) and interpeduncular nucleus-ventral tegmentum area (—— if significant; ... if nonsignificant). The two lines for glutamic acid decarboxylase did not differ significantly one from the other.*

than in cell body areas such as the ventral tegmentum *(McGeer et al., 1976)*.

C. CHANGES IN ENZYME LEVELS IN SENESCENT RATS

The level of TH in the whole brain, minus the striata and cerebellum, of Wistar rats is plotted against age in Figure 4. GAD, CAT and acetylcholinesterase (AChE) activities in the same brains showed rather similar trends with a peak activity at 7-15 months. Single factor analysis of variance of these data indicated no significant change across age for any of these enzymes in these samples *(McGeer, Fibiger, McGeer and Wickson, 1971)*.

In the same study, however, a significant change with age

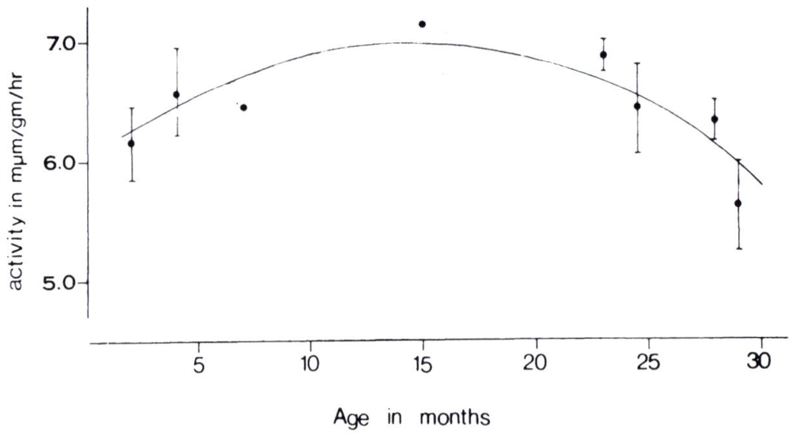

Figure 4. Tyrosine hydroxylase activity in whole rat brain minus cerebellum and neostriata as a function of age.

was found for TH, CAT and GAD in the striata (Figure 5). AChE activity was not significantly correlated with age even in the striata. It is clear that the change with age is most marked for TH.

Similar though less marked declines were noted in the TH activity of the caudate nucleus and olfactory tubercle in F 334 rats by Reis, Ross and Joh *(1977)*. These authors also found a small elevation in the activity of TH and a small reduction in DBH activity in the hypothalamus, a region containing both dopaminergic and noradrenergic terminals. There was no change in the activities of TH or DBH in the locus coeruleus or hippocampus which also are associated with noradrenergic neurons. These authors found no significant age changes in any of the enzymes in mice. McNamara, Miller, Benignus and Davis *(1977)* report data indicating less synthesis of dopa in 24 month old than in 9 month old rats and Algeri, Bonati, Brunello and Ponzio *(1977)* found a substantial decline in TH activity in the striatum, cortex, diencephalon and brainstem between 3 month and 30 month old Wistar rats. In the latter study, there was an increase in the activity of dihydropteridine reductase, an enzyme associated with the co-factor essential for catecholamine synthesis. Finch and his colleagues have reported extensive data indicating decreased dopaminergic activity in the brains of senescent mice *(Finch, 1973 and 1977; Jonec and Finch, 1975)*. They have considered that this decrease may have important interrelations with endocrine changes in aging. Various other laboratories have reported findings of decreased catecholaminergic activity in aged rats and some have similarly tried to involve changes in endocrine function *(Clemens and Bennett, 1977; Simpkins, Mueller, Huang and Meites, 1977;*

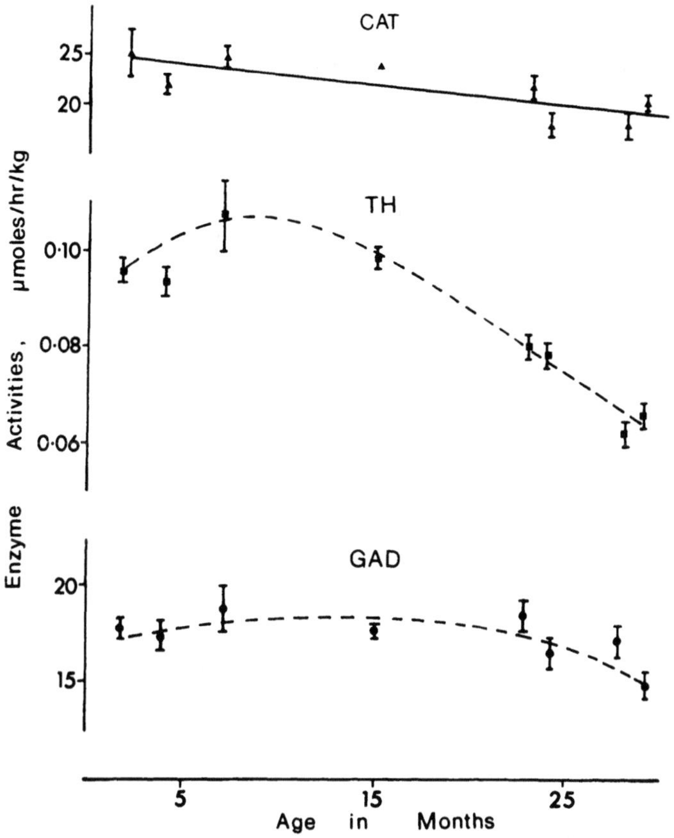

Figure 5. Tyrosine hydroxylase (TH), choline acetyltransferase (CAT) and glutamic acid decarboxylase (GAD) activities in rat neostriata as a function of age.

Miller, Shaar and Reigle, 1976; Sun, 1976).

These data generally support the concept of declining catecholamine activity in the brain of aging rodents with dopaminergic systems being more vulnerable than noradrenergic ones.

Losses with respect to cholinergic systems are of a more modest nature and are, therefore, of more questionable significance. Frolkis, Bezrukov, Duplenko, Shchegoleva, Shevtchuk and Verkhratsky *(1973)* report decreased activity of both CAT and AChE with age, while Meek, Bertilsson, Cheney, Zsilla and Costa *(1977)* indicate significant losses in the cholinergic system are restricted to the caudate nucleus. Hollander and Barrows *(1968)* found a minimal decrement from 3-24 months in forebrain (17%) and cerebral cortex (10%) AChE; while Kaur and Kanungo *(1970)* report a loss of about 25% in AChE activity in rat cerebral hemisphere from 20-96 weeks. The decrease on aging of CAT activity in rat reported by Verhratsky *(1970)* is roughly comparable to the decrease in caudate activity indicated in Figure 5.

Epstein and Barrows *(1969)* found no significant change with age from 2-26 months in GAD activity in the brainstem, cerebellum, and prosencephalon. These results are in general accord with our finding of no significant correlation with age of GAD in the whole brain minus the caudate and cerebellum. The caudate GAD decline observed in our study was marginal in nature.

These data in rats indicate neuronal losses in senescence which vary according to the region of the brain and the type of neuron. Most vulnerable appears to be tha dopaminergic nigrostriatal system. Cholinergic declines are less dramatic but apparently detectable while those for GAD are slight and perhaps not of significance.

STUDIES IN HUMANS

CATECHOLAMINE SYSTEMS

A sharp decline of TH activity in humans with age was found by us in the caudate, putamen and nucleus accumbens *(McGeer and McGeer, 1976a)*. It was the most striking example of age related enzyme changes found in our entire study. There were, however, eleven areas where five or more samples were analysed which showed no significant correlation between TH activity and age, although the trend always seemed to be downward. These areas were the hippocampus, anterior perforating substance, septal area, olfactory tubercle, anterior globus pallidus, hypothalamus, substantia nigra, locus coeruleus and the red nucleus. In view of the rat data on TH, and the striking effect in the human neostriatum, it seems likely that a significant decline would be found in all areas if a sufficient number of samples were analysed. The decline in TH with age is shown in Figure 6 for the caudate nucleus and putamen. The decreases were much more rapid for TH than for CAT or GAD. The most dramatic declines were between the ages of 5 and 15 when the cells in the substantia nigra begin to develop their pigment. Cote and Kremzner *(1974)* report declines with age in TH activity of the striatum and substantia nigra, but others *(Robinson, Sourkes, Nies, Harris, Spector, Bartlett and Kaye, 1977; Grote et al., 1974)* have found no statistically significant relation between TH and age in their series. A number of laboratories have reported significant losses in dopamine *(Carlsson and Winblad, 1976)* and noradrenaline *(Robinson et al., 1977)* with age in human brain tissue although Robinson *et al.(1977)* found no overall pattern for dopamine.

In view of the dramatic decline in TH levels with age, cell counts were carried out on a series of human brains *(McGeer, McGeer*

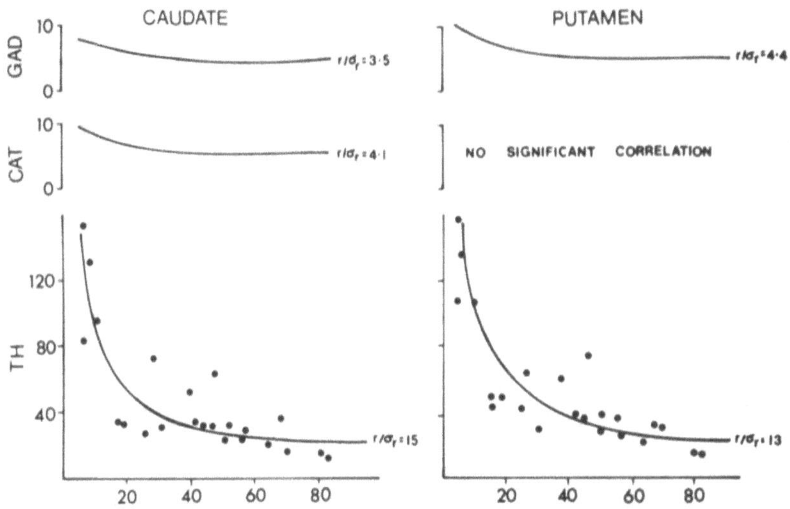

Figure 6. Calculated curves for tyrosine hydroxylase (TH), choline acetyltransferase (CAT) and glutamic acid decarboxylase (GAD) in the human caudate and putamen as a function of age. The TH in the nucleus accumbens showed a similar curve (TH = 22.5 + 466/age, r/σ_r = 14).

and Suzuki, 1977). The study showed that there was a significant decrease in the number of cells with age (Figure 7a). In neonatal brains about 400,000 dopaminergic neurons were counted in each substantia nigra. By age 60, the number had dropped to about 250,000. In Parkinson's disease (designated by the symbol p in the figure), cell counts ranged from 60,000 to 120,000. The symptoms of the disease are due to the loss of substantia nigra cells, but obviously many cells can disappear before decompensation occurs. The only stigmata may be the shuffling gait and stooped posture. often seen in the elderly. The loss in cells seems to be a slow and gradual decline, unlike the curvilinear drop seen in TH in the neostriatum. The two processes are obviously not directly related. It may be that there is a more rapid loss of nerve endings than cell bodies, or it may be that there is some kind of adjustment of TH levels unrelated to morphological factors. Synthetic enzymes such as TH are normally present in considerable excess in neurons.

Brody *(1976)* found a similar cell loss of noradrenergic cells in the locus coeruleus (Figure 7b). Here the numbers dropped from about 19,000 for young people to about 10,000 for people in their 80s. The calculated lines of regression are significant in both the locus coeruleus and the substantia nigra although the rate of loss is greater in the substantia nigra. As Brody points out,

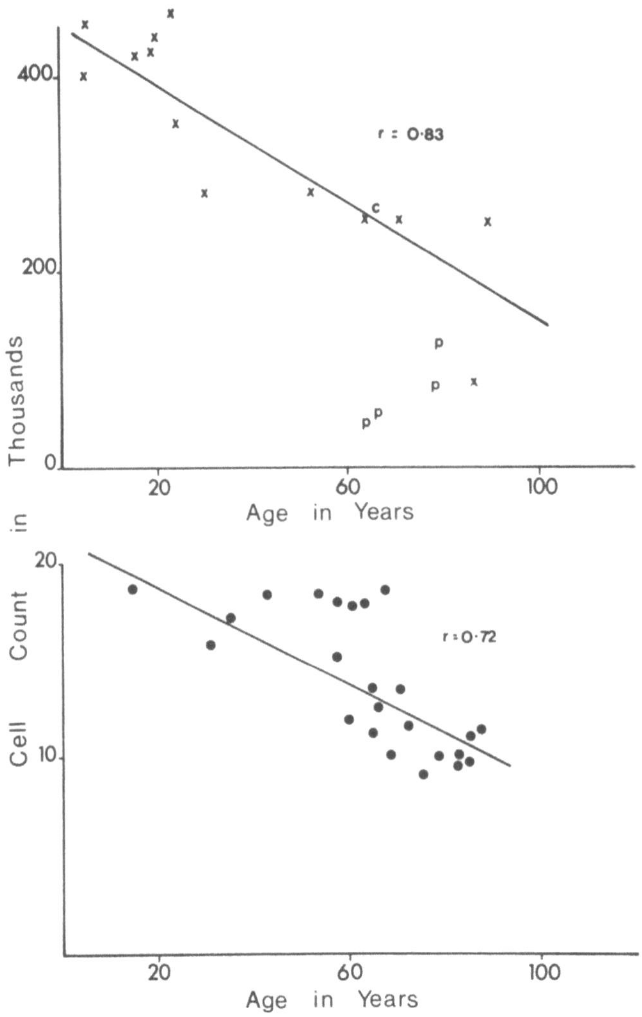

Figure 7. Calculated lines of regression for number of cells vs. age in the (a) human substantia nigra (top) and (b) locus coerulus (bottom). In the top graph, line and "x(s)" are for neurologically normal controls; "p" indicates a Parkinson's disease case and "c" a choreic.

such losses are not usual, for most brainstem nuclei so far studied showed no detectable decrements in cell number. A comparison of Figure 7a and 7b suggests that in humans, as in rats, the dopamine system seems to be more vulnerable than the noradrenergic one.

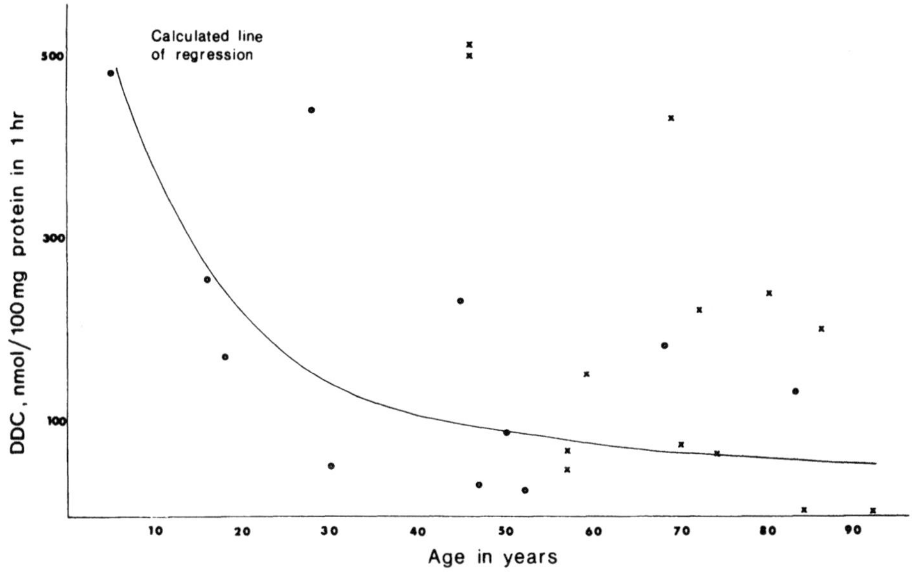

Figure 8. Dopa decarboxylase activity in the putamen as a function of age "o(s)" and line of regression are data of McGeer and McGeer (1976a) while "x(s)" are data of Lloyd and Hornykiewicz (1972).

DOPA DECARBOXYLASE (DDC)

As with TH, there were substantial declines of DDC with age in many areas (Figure 8), and these were very marked. Only a few areas such as the olfactory tubercle, preoptic area, anterior globus pallidus and hypothalamus showed no significant decrease of DDC with age, but all showed a definite trend. As indicated in Figure 8, the variability in DDC activities determined both by ourselves *(McGeer and McGeer, 1976a)* and by Lloyd and Hornykiewicz *(1972)* seems greater than the variability in TH activity.

DDC activity is found not only in catecholaminergic neurons but also in serotonergic neurons. It is difficult, therefore, to draw direct comparisons between changes in DDC activity and cell losses in the substantia nigra and locus coeruleus.

There is a very large excess of DDC in animal brains and presumably also in young humans so that age related declines in DDC are not apt to be as physiologically significant as declines in TH. The clinical benefits from dopa therapy in many cases of parkinsonism, a condition where DDC activity is abnormally low, support the view that the levels of decarboxylase activity even in aged brains are probably sufficient for production of the neurotransmitter amines.

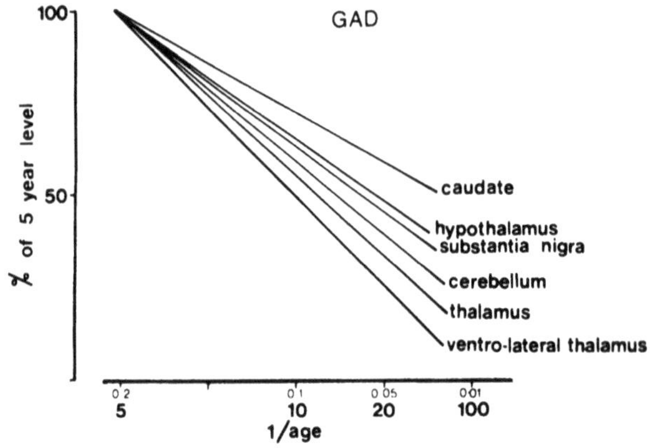

Figure 9. Representative curves for GAD activity as a function of age in various human brain regions. Activity vs age would give curves similar to those for CAT vs age in the caudate and g.p. (Figure 10). The significant linear correlations for GAD were with the reciprocal of age (light numbers on x axis); the corresponding ages are given by the dark numbers.

It is worth noting that several groups have reported an increase of MAO activity with age *(Cote and Kremzner, 1974; Grote et al., 1974; Robinson, 1975; Robinson, Nies, Davis, Bunney, Davis, Colburn, Bourne, Shaw and Coppen, 1972; Samorajski and Rolsten, 1973)*. Brain MAO is believed to exist in two forms (A and B). The picture is therefore rather complicated, and it is not certain whether the increase in MAO is related to neurons, glia, or both. It has been pointed out, however, that the presence of increased quantities of destroying enzymes, such as MAO, coupled with declining concentrations of synthetic enzymes such as TH and DDC would both work to reduce the amount of catecholamine transmitter available for physiological action.

GABANERGIC SYSTEMS

Glutamic acid decarboxylase (GAD), the enzyme responsible for the synthesis of GABA, is widely and rather evenly distributed in the grey matter. Of the fifty-six areas in our series where five or more samples were available, almost half showed a significant decline of GAD with age. The best fit for the data was almost always curvilinear indicating a more severe loss in the younger age groups (Figure 9). However the areas showing the most significant and greatest losses in GAD were not the same as those showing the greatest losses in TH or DDC. In general, the thalamic areas showed the greatest declines in GAD, followed by cortical and rhinencephalic areas. The basal ganglia showed relatively less decline of GAD with age. Perry, Gibson, Blessed, Perry and Tomlinson (*1977*) recently reported data indicating losses of GAD with age in many human brain areas.

CHOLINERGIC SYSTEMS

Acetylcholine is synthesized by CAT and metabolized by AChE. Although CAT is very highly concentrated in the caudate, putamen and parts of the limbic system, appreciable activity can be measured in every area of brain examined. Of those areas where five or more samples were available in our series, almost a third showed a significant correlation between CAT activity and age. In contrast with TH, DDC and GAD, the declines in CAT activity with age were particularly notable in cortical areas, while being less in extrapyramidal and rhinencephalic structures (Figure 10). Bird and Iverson (*1974*) found no decline of CAT in the putamen with age and decline in our series did not reach significance in the putamen although it did in the caudate. The decline of CAT with age in the

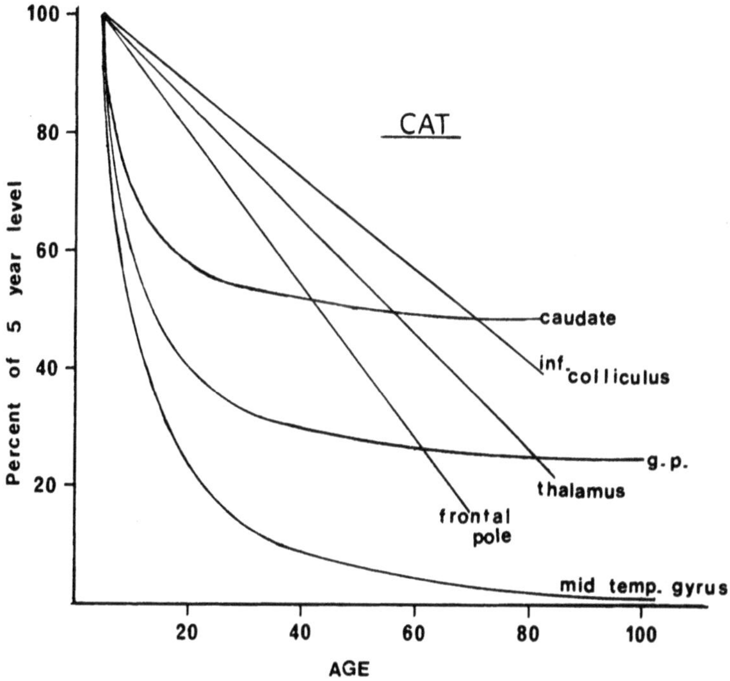

Figure 10. Representative curves for CAT activity as a function of age in various human brain regions.

cortex may be related to the dropout of cortical cells reported by Brody *(1976)*, since cholinergic cells have been found in the cortex by immunohistochemical staining *(McGeer, McGeer, Singh and Chase, 1974)*. At this stage that remains a speculation.

Losses in CAT activity, particularly in the cortex, have been reported in Alzheimer's disease and related disorders *(Bowen, White, Flack, Smith and Davison, 1974; Perry et al., 1977; Spillane, White Goodhardy, Flack, Bowen and Davison, 1973)*, as well as in normal aging *(Perry et al., 1977; Davies, 1978)*. Muscarinic cholinergic receptor binding is reportedly unaffected in Alzheimer's disease *(Davies and Verth, 1977)* suggesting that the deficiency may be limited to cholinergic neurons and not cholinoceptive neurons.

AChE activity generally parallels CAT activity. Somewhat fewer regions showed a significant correlation with age in AChE, but AChE, like MAO, is present in excess and is not limited to neuronal locations. Hence, it is not as satisfactory an index of neuronal vitality as is the synthetic enzyme.

DISCUSSION

Much more obviously needs to be done in defining the particular cellular and chemical losses which occur both in normal aging and in accelerated aging problems such as senile dementia. Only a few of the systems have been investigated so far, and the data are only preliminary in nature.

An even more fundamental question, however, that needs to be addressed is why these losses should occur at all. Why do particular kinds of cells in given regions die or lose their activity? There are many possibilities. It could be the gradual accumulation of insoluble debris, such as lipofuscin, which impairs the ability of the neurons to function. It could be a failure in some growth promoting factor or factors such as the nerve growth factor discovered by Levi-Montalcini and Cohen *(1960)* for peripheral synpathetic neurons and dorsal root ganglian neurons. It could be a declining ability of the axoplasmic transport system to sustain activity at nerve endings. It could be damage due to toxins, or even slow viruses. Or it could be any one of a number of other factors not suggested in this list.

Phenomena associated with aging are becoming of increasing interest in physiology and medicine. The conquest of specific disease processes inevitably leads to a larger percentage of the population being affected by age related disorders. In the case of aging, there is enormous cost to society of maintaining individuals whose mental and neurological capacities have slipped below the threshold necessary for independence. Therefore, it might be anticipated that there would be great practical benefits accruing from further research in this area.

REFERENCES

Algeri, S., Bonati, M., Brunello, N. and Ponzio. (1977). Dihydropteridine reductase and tyrosine hydroxylase activities in rat brain during development and senescence: a comparative study. *Brain Res. 132,* 569-574.

Bird, E.D. and Iversen, L.L.(1974). Huntington's chorea: post mortem measurement of glutamic acid decarboxylase, choline acetyltransferase and dopamine in basal ganglia. *Brain 97,* 457-472.

Bowen, D.M., White, P., Flack, R.H.A., Smith, C.B. and Davison, A. N. (1974). Brain decarboxylase activities as indices of pathological change in senile dementia. *Lancet i,* 1247-1249.

Brody, H. (1976). An examination of cerebral cortex and brainstem aging. In: *Aging*, vol. 3, pp. 177-181. Eds. R.D. Terry and S. Gershon. Raven Press, New York.

Carlsson, A. and Winblad, B. (1976). Influence of age and time interval between death and autopsy on dopamine and 3-methoxytyramine levels in human basal ganglia. *J. Neural Trans. 38,* 271-276.

Clemens, J.A. and Bennett, D.R. (1977). Do aging changes in the preoptic area contribute to loss of cyclic endocrine function? *J. Geront. 32,* 19-24.

Cote, L.J. and Kremzner, L.T. (1974). Changes in neurotransmitter systems with increasing age in human brain. *Trans. Am. Soc. Neurochem. 5,* 83.

Coyle, J.T. and Campochiaro, P. (1976). Ontogenesis of dopaminergic-cholinergic interactions in the rat striatum: a neurochemical study. *J. Neurochem. 27,* 673-678,

Davies, P. (1978). Studies on the neurochemistry of central cholinergic systems in Alzheimer's disease. In: *Alzheimer's Disease - Senile Dementia and Related Disorders.* Eds. R.D. Terry and R. Katzman, Raven Press, New York.

Davies, P. and Verth, A.H. (1977). Regional distribution of muscarinic acetylcholine receptor in normal and Alzheimer's-type dementia brains. *Brain Res. 138,* 385-392.

Epstein, M.H. and Barrows, C.H., Jr. (1969). The effects of age on the activity of glutamic acid decarboxylase in various regions of the brains of rats. *J. Geront. 24,* 136-139.

Finch, C.E. (1973). Catecholamine metabolism in the brains of aging male mice. *Brain Res. 52,* 261-276.

Finch, C.E. (1977). Neuroendocrine and autonomic aspects of aging. In: *Handbook of the Biology of Aging,* pp. 262-280. Eds. C. E. Finch and L. Hayflick, Van Nostrand Reinhold, New York.

Frolkis, V.V., Bezrukov, V.V., Duplenko, Y.K., Shcheguleva, I.V., Shevtchuk, V.G. and Verkhratsky, N.S. (1973). Acetylcholine metabolism and cholinergic regulation of functions in aging. *Gerontologia 19,* 45-57.

Grote, S.S., Moses, S.G., Robins, E., Hudgens, R.W. and Croninger, A.B. (1974). A study of selected catecholamines metabolizing

enzymes: A comparison of depressive suicides and alcoholic suicides with controls. *J. Neurochem.* *23*, 791-802.
Hattori, T. and McGeer,P.L. (1973). Synaptogenesis in the corpus striatum of infant rats. *Expl Neurol.* *38*, 70-79.
Hattori, T., Singh, V.K. McGeer, E.G. and McGeer P.L. (1976). Immunohistochemical localization of choline acetyltransferase containing neostriatal neurons and their relationship with dopaminergic synapses. *Brain Res.* *102*, 164-173.
Hollander, J. and Barrows, C.H., Jr. (1968). Enzymatic studies in senescent rodent brains. *J. Geront.* *23*, 174-179.
Jonec, V. and Finch, C.E. (1975). Aging and dopamine uptake by subcellular fractions of the C57BL/6J male mouse brain. *Brain Res.* *91*, 197-215.
Kaur, G. and Kanungo, M.S. (1970). Alterations in glutamate dehydrogenase of the brains of rats of various ages. *Can J. Biochem.* *48*, 203-206.
Levi-Montalcini, R. and Cohen, S. (1960). Effects of the extract of the mouse submaxillary salivary glands on the sympathetic system of mammals. *Ann. N.Y. Acad. Sci.* *85*, 324-341.
Lloyd, K.G. and Hornykiewicz, O. (1972). Occurrence and distribution of aromatic L-amino acid (L-DOPA) decarboxylase in the human brain. *J. Neurochem.* *19*, 1549-1559.
McGeer, E.G., Fibiger, H.C., McGeer, P.L. and Wickson, V. (1971). Aging and brain enzymes. *Expl. Geront.* *6*, 391-396.
McGeer, E.G., Fibiger, H.C. and Wickson, V. (1971). Differential development of caudate enzymes in neonatal rat. *Brain Res.* *32*, 433-440.
McGeer, E.G., Gibson, S., Wada, J.A. and McGeer, P.L. (1967). Distribution of tyrosine hydroxylase activity in adult and developing brain. *Can. J. Biochem.* *45*, 1943-1952.
McGeer, E.G. and McGeer, O.L. (1976a). Neurotransmitter metabolism in the aging brain. In: *Aging*, vol. 3, pp. 389-403. Eds. R.D. Terry and S. Gershon, Raven Press, New York.
McGeer, P.L. and McGeer, E.G. (1976b). Enzymes associated with the metabolism of catecholamines, acetylcholine and GABA in human controls and patients with Parkinson's disease and Huntington's chorea. *J. Neurochem.* *26*, 65-76.
McGeer, P.L., McGeer, E.G., Singh, V.K. and Chase, W.H. (1974). Choline acetyltransferase localization in the central nervous system by immunohistochemistry. *Brain Res.* *81*, 373-379.
McGeer, P.L., McGeer, E.G. and Suzuki, J.S. (1977). Aging and extrapyramidal function. *Archs. Neurol.* *34*, 33-35.
McGeer, E.G., Parkinson, J. and McGeer, P.L. (1976). Neonatal enzyme development in the interpeduncular nucleus and surrounding ventral tegmentum. *Expl. Neurol.* *53*, 109-114.
McNamara, M.C., Miller, A.T., Jr., Benignus, V.A. and Davis, J.N. (1977). Age related changes in the effect of electroconvulsive shock (ECS) on the *in vivo* hydroxylation of tyrosine and tryptophan in rat brain. *Brain Res.* *131*, 313-320.

Meek, J.L., Bertilsson, L. Cheney, D.L., Zsilla, G. and Costa, E. (1977). Aging-induced changes in acetylcholine and serotonin content of discrete brain nuclei. *J. Geront. 32*, 129-131.

Miller, A.E., Shaar, C.J. and Riegle, G.D. (1976). Aging effects on hypothalamic dopamine and norepinephrine content in the male rat. *Expl. Aging Res. 2*, 475-480.

Perry, E.K., Gibson, P.H., Blessed, G., Perry, R.H. and Tomlinson, B.E. (1977). Neurotransmitter enzyme abnormalities in senile dementia: choline acetyltransferase and glutamic acid decarboxylase in necropsy brain tissue. *J. Neurol. Sci. 34*, 247-265.

Reis, D.J., Ross, R.A. and Joh, T.H. (1977). Changes in the activity and amounts of enzymes synthesizing catecholamines and acetylcholine in brain, adrenal medulla, and sympathetic ganglia of aged rat and mouse. *Brain Res. 136*, 465-474.

Robinson, D.S. (1975). Changes in monoamine oxidase and monoamines with human development and aging. *Fed. Proc. 34*, 103-107.

Robinson, D.S., Nies, A., Davis, J.N., Bunney, W.E., Davis, J.M., Colburn, R.W., Bourne, H.R., Shaw, D.M. and Coppen, A.J. (1972). Aging, monoamines and monoamine oxidase levels. *Lancet i*, 290-291.

Robinson, D.S., Sourkes, R.L., Nies, A., Harris, L.S., Spector, S., Bartlett, D.L. and Kaye, I.S. (1977). Monoamine metabolism in human brain. *Arch. Gen. Psychiat. 34*, 89-92.

Samorajski, T., and Rolsten, C. (1973). Age and regional differences in the chemical composition of brain of mice, monkeys and humans. In: *Progress in Brain Research*, vol. 3, pp. 253-265. Ed. D.H. Ford, Elsevier Scientific Publishing Co., Amsterdam-London-New York.

Simpkins, J.W., Mueller, G.P., Huang, H.H. and Meites, J. (1977). Evidence for depressed catecholamine and enhanced serotonin metabolism in aging male rats: possible relation to gonadotropin secretion. *Endocrinology 100*, 1672-1678.

Spillane, J.A., White, P., Goodhardt, M.J., Flack, R.H.A., Bowen, D.M. and Davison, A.N. (1973). Selective vulnerability of neurons in organic dementia. *Nature 266*, 558-559.

Sun, A.Y. (1976). Aging and *in vivo* norepinephrine-uptake in mammalian brain. *Expl. Aging Res. 2*, 207-219.

Tennyson, V.M., Barrett, R.E., Cohen, G., Cote, L., Heikkila, R. and Mytilineou, C. (1973). Correlation of anatomical and biochemical development of the rabbit neostriatum. *Prog. Brain Res. 40*, 203-217.

Verkhratsky, N.S. (1970). Acetylcholine metabolism peculiarities in aging. *Expl. Geront. 5*, 49-56.

HETEROGENEITY OF POLYPEPTIDE HORMONES DURING AGING

THOMAS L. KLUG, MARK F. OBENRADER, and RICHARD C. ADELMAN

Fels Research Institute, Temple University School of Medicine, Temple University Institute on Aging and Division of Biomedical Research, Philadelphia Geriatric Center, Philadelphia, Pennsylvania 19140

INTRODUCTION

Alterations in the structure of DNA, RNA and protein molecules are a well established biochemical manifestation of aging. These structural changes were originally detected as alterations in physical properties such as stability to heat denaturation or changes in intrinsic biological activity. However, these studies have not progressed to the point where a specific structural alteration of a known type occurring in a macromolecule may be held responsible for the decreased functional capacity characteristic of senescence. This latter condition if not surprising in view of (1) the tremendous structural complexity and functional interdependence of DNA, RNA and protein structures, and (2) the difficulty of correlating specific, frequently small, alterations in molecular structure with changes in biological and ultimately physiological activities.

Within the current framework of knowledge and analytical techniques available for the study of molecular biology, alterations in the structure of proteins seem to be the most accessible to immediate study. Recent advances in polyacrylamide electrophroesis, isoelectric focusing and radioimmunoassay methods in particular allow detection of the smallest of perturbations in the fine structure of proteins. The former methods permit the investigator to detect the presence of different forms of a single protein due to differences in molecular weight, charge, conformation and in certain situations primary sequence. Since proteins are the primary expression of genetic information, age-dependent alterations in the structure or organization of DNA and RNA should be expressed in at least the quantitative and perhaps qualitative pattern or structural nature of the proteins synthesized during aging. The potential difficulties inherent in this approach to the problem of aging are perhaps best illustrated by an example. Consider a protein X which possesses a biological effect or activity Y, and presume that an altered form of X,X', accumulates during aging. Firstly, does the presence of X' result from altered transcription, translation, posttranslational or metabolic processes, or some combination of these? If altered transcriptional processes are implicated, then does the altered protein X' accumulate as a direct consequence, or only indirectly through effects of the altered transcription on other proteins involved in translation and posttranslational modifications? Secondly, what is the biological activity of the altered protein X'? Does X' express new, perhaps deleterious effects, and exactly what structural features of X' are responsible for these effects? While these questions may seem formidable, the very indeterminancy of this approach is its chief advantage, as it encourages a mechanistic approach to studies in biochemical aging - an approach essentially free of experimental bias. Experiments to determine the type and nature of the structural modification(s) yielding the altered protein would partly determine the organizational level at which the altered protein first appeared, e.g., an alteration in the amino acid primary sequence would imply altered transcription or translation. These results would also suggest experiments to determine the mechanism responsible for the appearance of the altered protein during senescence. The problem of assessing the biological consequences of the altered proteins may be met by studying a polypeptide with a highly specific and readily assayable biological activity. In this regard, the enzymes and polypeptide hormones seem the most appropriate since the activity of these proteins are highly sensitive to perturbations in their structure.

In this article we will present some of our initial experimental results on the presence during aging of altered or "heterogeneous" forms of two very different polypeptide hormones; insulin and thyrotropin (TSH). Because the complexity and nature of structural variability in polypeptide hormones may not be generally appreciated, the source and methods of detecting heterogeneous forms of polypeptides will be discussed in some detail. In this regard, those aspects of polypeptide heterogeneity relating to insulin and TSH structure will be particularly emphasized. For a more comprehensive discussion of the heterogeneity of other polypeptide hormones, the reader is referred to several reviews of the subject *(Rabinowitz and Roth, 1974; Yalow, 1974)*. The evidence for alterations in the heterogeneity of insulin and TSH during aging, and a discussion of its physiological and clinical significance will be presented. Finally, the implications of polypeptide heterogeneity with respect to certain current theories of aging will be discussed.

SOURCES AND DETECTION OF POLYPEPTIDE HETEROGENEITY

Polypeptide hormones were originally isolated and purified by chemical and physical methods which used criteria of biological activity to develop separation techniques and assess the purity of the hormones obtained. These preparations were, therefore, necessarily homogeneous with respect to the form of the hormone isolated. With the development of the radioimmunoassay (RIA) technique using antisera to purified ACTH and insulin, Berson and Yalow demonstrated by gel filtration chromatography that serum and glandular tissue extracts contained several hormonal forms of apparently different molecular weights *(Yalow, 1974)*. These immunologically cross-reactive heterogeneous forms of the polypeptide hormones may be due to the presence of one or more of the following: (1) high molecular weight biosynthetic precursors; (2) high molecular weight (degradative) metabolites; (3) metabolic fragments of the polypeptide hormone; (4) structural alterations in polypeptide amino acid side chains; and (5) polypeptide pleiomorphism.

Large molecular forms of polypeptide hormones which may function as biosynthetic precursors have been identified for ACTH *(Yalow and Berson, 1970)*, insulin *(Steiner, Cunningham, Spiegelman and Aten, 1967)*, parathyroid hormone *(Sherwood, Rodman and Lundberg, 1970)*, glucagon *(Noe and Bauer, 1971)*, gastrin *(Yalow and Berson, 1970)*, human placental lactogen *(Friesen, Guyda and Hardy, 1970)*, calcitonin *(Roos, Okano and Deftos, 1974)*, and growth hormone *(Friesen et al., 1970)*. There is also partial evidence to support the existence of high molecular weight intermediates in the biosynthesis of pituitary glycoproteins such as follitropin (FSH) *(Reichert and Ramsey, 1977)*, lutropin (LH) *Prentice and Ryan, 1975)* and TSH *(Klug and Adelman, 1977; Erhardt and Scriba, 1977)*. The presence of these precursorial forms of the hormones was initially demonstrated by gel filtration chromatography and radioimmunoassay of glandular tissue extracts. More recently, cell-free translation of RNA extracts of glandular origin has been used to demonstrate these precursors or prohormones and previously undetected precursorial forms (Preprohormones) *(Yip, Hew and Hsu, 1975; Kemper, Habener, Mulligan, Potts and Rich, 1974)*. While high molecular weight immunologically cross-reactive forms of insulin, growth hormone, and TSH have reportedly been found in serum, their identification as biosynthetic precursors normally found in glandular tissue extracts is less easily demonstrated, as postsecretory factors may lead to the formation of species with similar behavior.

Antoniades *(1975;1976)* found that in the rat intact forms of (^{125}I) growth hormone were converted into high molecular weight forms following intravenous injection. Denaturation with 8M urea and 2% SDS did not dissociate these immunologically cross-reactive species, suggesting that their formation may involve covalent binding. When denaturation was combined with sulfitolysis, high molecular weight (^{125}I) insulin was reduced to free A- and B-chains, implying that disulphide binding may be involved. The formation of these species reportedly did not occur in serum alone. Although their potential significance was not elucidated, these previous studies also revealed the existence of immunoreactive fragments of the injected radiolabeled hormones.

Immunochemical heterogeneity of a polypeptide hormone due to the presence of peripherally produced metabolites was first demonstrated for human parathyroid hormone *(Berson and Yalow, 1968)*. Early studies of this hormone in plasma could not resolve the extent to which this immunochemical heterogeneity was due to the presence of peripherally produced metabolites or glandular secretion of different hormonal forms. Studies done since, however, utilizing radioimmunoassays with antisera with different immunological specificities have demonstrated that steady-state plasma concentrations of immunoreactive material probably result form both peripheral conversion and glandular secretion of several forms of parathyroid hormone *(Silverman and Yalow, 1973; Reiss and Canterbury, 1974)*. The characterization of immunoreactive molecular species as simply polypeptide hormone or fragments of this hormone is not always possible, since the fragments of one hormone may be identical to the intact structure of a smaller hormone. Consider, for example, the two different melanocyte-expanding principles, α- and β-MSH. The peptide chain of α-MSH is identical with that of adrenocorticotropin (ACTH) in the N-terminal 13 amino acid residues. Similarly, the β-MSH sequence is contained within the 19 N-terminal amino acid residues of lipotropic hormone (γ-LPH) *(Cretien and Li, 1967)*. In addition, β-LPH similarly contains the sequences for β-endorphin and the enkephalins. The possibility that the larger molecules ACTH and β-LPH may serve as the biosynthetic precursors for α- and β-MSH, and β-endorphin remains to be unequivocally demonstrated. An even larger common precursor to both ACTH and β-LPH was recently reported, making the relationship between all of the former polypeptide hormones very complex *(Roberts and Herbert, 1978)*. Another sort of identity crisis may occur for the pituitary glycoprotein hormones TSH, LH and FSH. These hormones are composed of two nonidentical non-covalently bound subunits, α and β. The β subunit confers both the biological and immunological specificity; the α subunits of these hormones are very similar, but not necessarily identical in structure *(Vaitukaitis and Ross, 1974)*. The separate subunits may be secreted separately and recombine such that any type of α subunit may recombine with, e.g., a TSH-β subunit to give the appropriate TSH biological activity *(Pierce, Bahl, Conwell and Swaminathan, 1971)*.

These previous sources of polypeptide heterogeneity may be yet further complicated by pleiomorphism, i.e., multiple forms of a single polypeptide hormone which differ in primary sequence. The existence in the fish and rat of at least two forms of insulin (I and II) is a very good example of this kind of heterogeneity

(Smith, 1964). Rat insulins I and II differ by a single nonconservative amino acid substitution, a lysine (insulin I) for a methionine (insulin II) at position 29 of the B-chain *(Clark and Steiner, 1969).* This nonconservative substitution results in a net charge difference which forms the basis for separation on DEAE cellulose. In a situation where a single conservative amino acid substitution exists, this pleiomorphic form of the hormone might easily be overlooked by present physical and chemical separation techniques, unless one is specifically looking for such hormones. The two rat insulins are nonallelic (homozygous), i.e., coded for by two genes *(Smith, 1966).* Another pancreatic polypeptide hormone, glucagon, shows considerable sequence homology and hence similar immunological behavior to glucagon-like peptides of gastrointestinal origin *(Faloona and Unger, 1974).* The homology between pancreato- and enteroglucagon-like peptides is so great that it has thus far proven impossible to prepare antibody with high specificity for only one form. In this case, however, the hormones may be considered as essentially distinct, as sequence differences probably reside between amino acids 22 and 27 *(Assan and Slusher, 1972).* A structural variant of human growth hormone has recently been isolated and partially characterized *(Lewis, Dunn, Bonewald, Seavey and Vanderlaan, 1978).* This form of growth hormone has a lower molecular weight than the predominant form of the hormone, contains sequence differences, but still possesses the full activity of growth hormone. These studies point out the potential confusion likely to occur in the discrimination between pleiomorphic forms of a single hormone and related, but unique polypeptides. Glycoproteins of pituitary origin, particularly TSH, may be disposed to polymorphism of this type. Several studies have reported possible sequence inversions and/or amino acid substitutions near the C-terminus of bovine and porcine LH-α subunit *(Papkoff, Sairam and Li, 1971).* The subunits of TSH-β also appear to be polymorphic by virtue of amide content *(Davy, Fawcett and Morris, 1977)*, carbohydrate, and conformational differences *(Guidice and Pierce, 1977; Pierce, 1974).*

Another potential source of polypeptide heterogeneity may be due to the presence of heterogeneous cell populations in the endocrine tissues. These different endocrine cell populations may secrete slightly different forms of the respective polypeptide hormone, with the secretion of each cell type being under the control of separate secretagogues. As an example, we have recently reported a differential insulin secretory response to glucose stimulation in large vs small islets of Langerhans in the male rat (manuscript submitted). Similar functional heterogeneity in somatotrophs *(Snyder, Hymer and Snyder, 1977)* and thyrothrophs *(Leuschen, Tobin and Moriarity, 1978)* isolated from the anterior pituitary of the rat was recently reported.

EVIDENCE FOR ALTERATIONS IN POLYPEPTIDE HETEROGENEITY DURING AGING

Preliminary experiments in our laboratories to assess the effects of aging on polypeptide hormone heterogeneity have centered upon two very different types of hormones: insulin and thyrotropin (TSH). The majority of these studies were done with barrier-maintained 3-, 12-, and 24-month old male Sprague-Dawley rats. The rearing, maintenance, and age-related pathology of these rats were described previously *(Cohen, Anver, Ringler and Adelman, 1978)*.

The pattern of glucose-stimulated secretion of insulin into portal vein blood was found to be progressively altered between 2- and 24-months of age in the male rat *(Gold, Karoly, Freeman and Adelman, 1976)*. The physiological significance of this difference is difficult to assess since the radioimmunoassay procedure alone cannot distinguish between the various active and inactive forms of insulin. We, therefore, examined the effect of aging on the heterogeneity (molecular weight distribution) of immunoassayable serum insulin following glucose refeeding *(Obenrader, Auth, Gold, Ceci, Kitahara and Adelman, 1978)*.

The apparent molecular weight distribution of immunoreactive insulin (IRI) recovered from acid-ethanol extracted portal vein serum was determined by gel filtration of the extracts with Sephadex G-50 equilibrated with 1N acetic acid. Fractions from the chromatographic procedure were collected, concentrated and assayed for the presence of insulin-like material by radioimmunoassay. This procedure was repeated for extracts of serum collected at hourly intervals of up to 7 hours following glucose refeeding. Several distinct peaks or components of IRI material, arbitrarily designated I-VIII, were detected at elution volumes corresponding to molecular weights greater and less than intact insulin: Fractions I and II eluted at positions indicating an apparent molecular weight greater than insulin; Fractions III and IV eluted at positions identical to rat proinsulin and insulin standards, respectively; Fractions V-VIII eluted at positions indicating a molecular weight less than insulin. The temporal pattern of IRI in each of the fractions following glucose refeeding was varied: Fractions I and II were maximal at 5 to 7 hours; Fractions III and IV varied in a pulsatile manner; Fractions V-VIII also exhibited a pulsatile time course over the 1 to 7 hours following glucose refeeding. While the temporal patterns of Fractions I-IV were similar for both the 2- and 24-month old rats, the total level of IRI was greater in the old rats. However, both the temporal pattern and magnitude of the apparently low-molecular-weight IRI differed considerably between 2- and 24-months of age. All of these insulin-like species are believed to be insulin or insulin metabolites, or molecules with extremely similar primary sequence to insulin, as the antisera used in these studies did not show any crossreactivity with glucagon, angiotensins, pancreozymin,

ACTH, growth hormone, secretin or TSH. Furthermore, this particular antiserum exhibited identical crossreactivity with porcine, bovine, human and rat insulin, indicating that the antigenic determinants recognized by the predominant classes of antiinsulin immunoglobulins in this antiserum are specific for a species invariable region of the insulin molecule.

Therefore, insulin in serum demonstrates a very complex instance of polypeptide heterogeneity, the sources of which are probably diverse. The apparently high molecular weight IRI material (Fractions I and II) may represent self-aggregation phenomena *(Permutt, Biesbroeck and Chyn, 1977)*, protein-protein disulphide exchange products *(Varandani, Shroyer and Nafz, 1972)*, or a large biosynthetic precursor *(Chan, Keim and Steiner, 1976; Permutt and Routmann, 1977)*. Fraction III probably is a proinsulin-like material consisting of several components *(Steiner, Holland, Rubenstein, Cho and Bayliss, 1969)*. Fraction III may also be related to the presence of the serum polypeptide hormone insulin-like growth factor I (IGF-I), a polypeptide exhibiting sequence homologies with proinsulin *(Rinderknecht and Humbel, 1978)*. The lower molecular weight IRI (Fractions V-VIII) could occur by several mechanisms. They could represent previously unidentified degradation products, perhaps due to postsecretory enzymatic cleavage by glutathione-insulin transhydrogenase *(Varandani, 1974)*, an insulin specific protease (ISP) *(Brush, 1971)*, or other enzymes. Premature termination or altered initiation of insulin peptides in response to glucose occurs in *in vitro* studies *(Permutt, 1974)*. Further studies are needed, however, to establish by independent methods whether these apparent low molecular weight IRI species are really fragments of insulin, or otherwise modified forms of insulin. For instance, the basic C-peptide fragment of proinsulin has an apparent molecular weight greater than insulin by gel filtration chromatography. These initial studies have ignored the potential contribution of polypeptide pleiomorphism (insulins I and II) to age-dependent differences in insulin heterogeneity.

The initial observation of an age-dependent decrease in the levels of bioassayable serum TSH while immunoassayable levels of TSH remained unchanged between the ages of 2- and 24-months in the male rat *(Klug and Adelman, 1977)*, originally led us to investigate the role of TSH heterogeneity during aging. The molecular weight distribution of immunoassayable TSH associated with serum from 2- to 24-month old rats was determined by fractionation of whole serum with Sephadex G-200 equilibrated with phosphosaline buffer. The presence of immunoreactive TSH (IR-TSH) or TSH-like material was detected by radioimmunoassay of aliquots from the various chromatographic fractions using a highly specific homologous rat radioimmunoassay (NIAMDD rat TSH RIA kit). Three areas of immunoreactivity were evident: V_e/V_o (I) 0.95-1.05; (II) 1.20-1.40; (III) 1.60-1.90. The serum from both 2- and 24-month old rats was characterized by a predominant immunoreactive peak which migrated at V_e/V_o 1.70, corresponding to intact pituitary TSH. Serum from

24-month old rats demonstrated greater amounts of an apparently high molecular weight immunologically crossreactive species. Upon rechromatography, this serum component migrated as a single peak at the identical elution volume. The contribution of these heterogeneous TSH-like species to serum bioassayable TSH activity was determined by bioassay of pooled serum fractions following fractionation by gel filtration with Sephadex G-200. The intact TSH peak, V_e/V_o 1.7, was found to be responsible for the thyroid-stimulating properties of the serum hormone. The high molecular weight elution peak, V_e/V_o approximately 1.0, corresponded to a thyroid-inhibitory activity, e.g., concentrated fractions containing the high molecular weight species lowered the levels of isotopically labeled thyroid hormones in intact assay mice below those of a control injection. This material also prevented the physiological response to exogenously injected TSH. Bioassay of mixed old and young serum also indicated that old serum contained greater amounts of an inhibitor of TSH biological activity *(Klug and Adelman, 1977)*.

The above results naturally led us to experiments to determine the nature of the high molecular weight TSH-like serum component; is it a TSH precursor, a TSH-carrier protein complex, or perhaps a TSH aggregate? At least one preliminary series of experiments with radioactively labeled pituitary extracts suggested that these may be TSH biosynthetic precursors. For example, the molecular weight distribution of immunoprecipitable TSH radioactivity in pituitaries from 24-month old rats following incubation with (^{3}H)-leucine showed an apparent heterogeneity similar to that of TSH in serum *(Klug and Adelman, 1977)*. Studies to further characterize these TSH-like molecules are currently in progress in our laboratories. Several other isolated studies have suggested that high molecular weight, perhaps biosynthetic precursors of TSH exist. An immunoreactive high-molecular-weight form of TSH was found in several pituitary extracts from patients with asymptomatic thyroiditis by Vanhaelst and Golstein-Golaire *(1976)*. These same authors reported previously the presence of a high molecular weight species in human serum that crossreacted with antisera to TSH-β subunit *(Golstein-Golaire and Vanhaelst, 1975)*. More recently, Erhardt and Scriba *(1977)* reported the isolation and partial characterization of a "big" TSH from human pituitary extracts. Other studies, like those of Lee, Aloj, Beguinot and Kohn *(1977)*, suggesting that serum contains a soluble TSH binding protein, may imply that more than one mechanism may be responsible for the apparent pattern of TSH heterogeneity in serum.

PHYSIOLOGICAL AND CLINICAL IMPLICATIONS

The potential physiological significance of the heterogeneous forms of insulin and TSH depends upon the ability of the various immunoreactive species to regulate the activity of the hormone-sensitive axis by (1) direct competition with the active hormone molecule in various target cell populations, or (2) modulating the levels of the active hormone available to the target cells. In the case of insulin, the nonspecific nature of the insulin bioassays dictates a prerequisite purification of each immunoreactive species before these possibilities can be investigated. Eventual studies of this nature may help elucidate those mechanisms which underlie glucose intolerance and insensitivity to insulin, as well as age-dependent changes in the regulation of insulin-sensitive enzymes such as hepatic glucokinase *(Gold et al., 1976; Freeman, Karoly and Adelman, 1973; Adelman and Freeman, 1972; Adelman, 1970)*. In the case of TSH, however, for which there are specific *in vivo* bioassays, at least one heterogeneous form of TSH was found to actually inhibit *in vivo* the thyroid-stimulating activity of endogenous and exogenous TSH *(Klug and Adelman, 1977)*. This latter finding may explain the etiology of certain hypothyroid or myxedematous states, such as the latent hypothyroidism which normally accompanies the aging process.

Previous studies of two other pituitary hormones, LH and FSH, demonstrated that the molecular forms of these hormones change in response to gonadal steroid feedback *(Diebel, Yamamoto and Bogdanove, 1973; Peckham, Yamaji, Dierschke and Knobil, 1973)*. The molecular form of TSH may then change during aging due to changes in the levels of the thyroid hormones, or alteration in the regulation of pituitary TSH by these hormones or TRH, as previously demonstrated in the aging male rat (manuscript submitted). In this regard, the various stages of prohormone to hormone conversion represent logical points for rapid physiological control of polypeptide hormone synthesis and could also serve as loci where aging or pathological processes could operate to influence the rate of hormone synthesis and/or the molecular form of the hormone secreted into the circulation.

The above findings also have broad implications for the use of routinely employed clinical radioimmunoassays of polypeptide hormones. If the radioimmunoassay is not specifically designed to distinguish between the various hormonally active, inert, or perhaps inhibitory heterogeneous forms, then such an assay would, in certain situations, give frankly false indications of physiological status. Selection or generation of antisera specific for the separate heterogeneous hormonal forms may appreciably assist in the detection of specific disease states. Another approach to the radioimmunoassay of the species may be to use purified preparations of these heterogeneous forms as radiolabeled ligands with crossreactive antisera. In this manner a single antiserum could be used to assay specifically for each of the heterogeneous forms.

PURSUIT OF MECHANISMS OF BIOCHEMICAL AGING

Age-dependent accumulation of altered proteins is not a new concept. While such altered or abnormal proteins have at one time or another been proposed to result from such diverse phenomena as the accumulation of free radical induced intermolecular cross-links or altered gene expression during senescence, only in the case of certain enzymes has the existence of these altered proteins been unequivocally established. Since the report of immunologically crossreactive but biologically inactive forms of liver aldolase in old mice by Gershon and Gershon *(1973)*, similar phenomena have been reviewed by Rothstein *(1977;1975)*. Our preliminary findings of qualitatively similar types of results for two polypeptide hormones may suggest that a common senescence-related mechanism may be operative for polypeptides in general. The analogy should not be extended too far, since alterations in the properties of the polypeptide hormones may be due to extracellular, e.g., postsecretory modifications, in addition to intracellular processes. It may not, however, be premature to speculate upon the possibilities which are: (1) failure to complete posttranslation modifications such as conversion of biosynthetic precursors to appropriate products, deamidation, disulphide reductions, oxidations, methylation, adenylation, carbohydration, phosphorylation, etc.; (2) conversion of the polypeptide into a partially degraded or conformationally alterted form: (3) release of incomplete polypeptides, or the substitution of individual amino acids (error theory); and (4) selective gene expression by selective gene transcription of translation. These categories are not necessarily mutually exclusive, since, for example, selective gene expression alone could easily originate all of the other possibilities.

Little evidence exists to support any particular explanation for the accumulation of altered enzymes or polypeptide hormones. Age-dependent changes in the electrophoretic properties of several enzymes isolated from human and bovine eye lens and old human erythrocytes may be due to posttranslational deamidation of asparagine and glutamine residues *(Skala-Rubinson, Vibert and Dreyfus, 1976; Eaton, Brewer and Tashin, 1966; Rosa and Shapira, 1964)*. Similar changes in electrophoretic properties of enzymes from tissues capable of *de novo* protein synthesis may be rare, since of all the altered enzymes studied *(Rothstein, 1977)*, only rabbit muscle aldolase exhibited an age-dependent electrophoretic heterogeneity attributable to deamidation *(Koida, Lai and Horecker, 1970)*. There is no evidence for differences in amidation state of polypeptide hormones during aging, although electrophoretic and isolectric focusing studies demonstrate that amidation state is a source of heterogeneity for insulin *(Steiner, Clark, Nolan, Rubenstein, Margoliash, Aten and Oyer, 1969)* and TSH*(Davy et al., 1977)*. The error theory *(Orgel, 1963)* has received little support from the extensive studies of viral replication in aging WI-38 cells *(Holland, Kohne and Doyle, 1973; Tompkins, Stanbridge and Hayflick, 1974)*, or studies of altered enzymes *(Rothstein, 1977)*.

The topic of selective gene expression deserves special consideration in the discussion of polypeptide hormones and aging. The operon mechanism which controls the kinds of products a prokaryotic cell produces through selective repression and expression of specific gene transcription may be applied to programmed gene expression in eukaryotes *(Jacob and Monod, 1963)*. Thus, sequential and selective gene expression may be involved in the mechanisms of aging and development. Translational control of development and aging may also be exerted through restricted translation of specific messages at approptiate periods in the lifespan *(Strehler, 1977)*. If either or both of these mechanisms are indeed operative, then one may reasonably expect to find alterations in the kinds and structural nature of the proteins synthesized during aging. These mechanisms assume particular importance for the polypeptide hormones, since many of these hormones are known to influence cellular differentiation, metabolic activity, and developmental events. Therefore, these polypeptides may be related to the products of so-called regulator genes *(Britten and Davidson, 1969)*, and changes in the heterogeneity of polypeptide hormones may be a primary, perhaps programmed event during senescence. For example, altered selective transcriptional regulation of the nonallelic insulin pleiomorph genes (rat insulins I and II) might lead to age-dependent changes in the predominant *in vivo* form of insulin during various phases of development and aging, resulting in changes in the function of insulin-sensitive tissues. Altered translation control may lead to similar results, but would also probably result in the synthesis of incomplete peptides in analogy to the consequences of point mutations, e.g., the production of mis-sense or nonsense proteins.

Perhaps the central thesis to recognize in the above discussion is that any theory of aging necessarily implies the existence and accumulation of altered proteins. The nature of the structural alteration in turn depends upon the mechanism responsible for its occurrence. Thus, the study of altered proteins provides an opportunity and method to discover the fundamental biochemical mechanisms of the aging process. While this experimental approach is relatively free of methodological bias, it still presents several experimental problems. It is not clear, for instance that all proteins exhibit structural heterogeneity, or that the same protein will exist in similar forms in different tissues, or in fixed vs clonally replicating cells. Therefore, initial searches for suitable altered protein systems in aging may be somewhat blind. There is some hope that direction for aging studies in this regard may be given by the study of specific disease states that may be related to the aging phenotype, e.g., the 'Segmental Progeroid Syndromes', as reviewed by Martin *(1978)*, leading one to suspect that alterations in some aspect of polypeptide hormone structure or metabolism may be present. Another problem in the detection and large scale isolation of the altered forms of a polypeptide is presented by the possibility that the physical and chemical properties of the 'abnormal' protein may either be the same as or very different from the 'normal' protein. In the former instance, a structural change such as a conservative amino acid substitution may be undetected. In the latter case, it may be difficult to demonstrate that the 'abnormal' protein is even related to the 'normal' protein without extensive studies. However, within this spectrum of possibilities, there are certainly many types of structural variations which are amenable to exploitation by current experimental techniques. The expectation for future progress in unraveling the biochemical mechanisms leading to the occurrence of altered proteins during aging may be influenced by the nature of these mechanisms. For example, if the altered proteins result from random error or 'noise' accumulation in cellular processes during aging, then the number and kinds of structural alterations in any single polypeptide may be so large as to prevent characterization by present analytical methods. If the altered protein results from either switching-off of specific genes involved in epigenetics *(Caplan and Ordahl, 1978)*, or switching-on of deleterious or late-acting genes, then the structural alterations are likely to be relatively limited in number and, therefore, more easily studied. Leaving questions of future progress aside, the simple distinction between these former alternative mechanisms for the appearance of altered proteins during aging would in itself represent a significant advance in gerontological research.

ACKNOWLEDGEMENTS

Dr. Richard C. Adelman is an Established Investigator of the American Heart Association. Support by the following research grants from the NIH also gratefully acknowledged: AG-00368 and CA-12227.

REFERENCES

Adelman, R.C. (1970). An age-dependent modification of enzyme regulation. *J. Biol. Chem. 245*, 1032-1035.

Adelman, R.C. (1973). Hormonal regulation of macromolecular synthesis during aging. In: *Mechanismes du vieillissement moleculaire et cellulare*, pp. 141-152. Paris: Les Colloques de L'Institut National de la Sante et de la Recherche Medicale.

Adelman, R.C. and Freeman, C. (1972). Age-dependent regulation of glucokinase and tyrosine aminotransferase activities of of rat liver *in vivo* by adrenal, pancreatic, and pituitary hormones. *Endocrinology 90*, 1551-1560.

Antoniades, H.N. (1975). Conversion of (^{125}I) growth hormone into high molecular weight forms *in vivo*. *Endocrinology 96*, 799-802.

Antoniades, H.N. (1976). Metabolism of single-component and high-molecular-weight radioactive insulin in rats. *Endocrinology 99*, 481-489.

Assan, R. and Slusher, N. (1972). Structure/function and structure/immunoreactivity relationships of the glucagon molecule and related synthetic peptides. *Diabetes 21*, 843-855.

Berson, S.A. and Yalow, R. (1968). Immunochemical heterogeneity of parathyroid hormone in plasma. *J. clin. Endocr. Metab.* *28*, 1037-1047,

Britten, R.J. and Davidson, E.H. (1969). Gene regulation for higher cells: a theory. *Science 165*, 349-357.

Brush, J.S. (1971). Purification and characterization of a protease with specificity for insulin from rat muscle. *Diabetes 20*, 140-155.

Caplan, A.I. and Ordahl, C.P. (1978). Irreversible gene repression model for control of development. *Science 201*, 120-130.

Chan, S.J., Keim, P. and Steiner, D.F. (1976). Cell-free synthesis of rat preproinsulins: characterization and partial amino acid sequence determination. *Proc. natn. Acad. Sci. USA 73*, 1964-1968.

Chretien, M. and Li, C.H. (1967). Isolation, purification, and characterization of γ-lipotropin hormone from sheep pituitary glands. *Can. J. Biochem. 45*, 1163-1174.

Clark, J.L. and Steiner, D.F. (1969). Insulin biosynthesis in the rat: demonstration of two proinsulins. *Proc. natn. Acad. Sci. 62*, 278-285.

Cohen, B.J., Anver, M.R., Ringler, D.H. and Adelman, R.C. (1978). Age-associated pathological changes in the male rat. *Fed. Proc.* (in press).

Davy, K.M.M., Fawcett, J.S. and Morris, C.J.O.R. (1977). Chemical differences between thyrotropin isohormones. *Biochem. J. 167*, 279-280.

Diebel, N.D., Yamamoto, M. and Bogdanove, E.M. (1973). Discrepancies between radioimmunoassays and bioassay in rat FSH: Evidence that androgen treatment and withdrawal can alter bioassay-immunoassay ratios. *Endocrinology 92*, 1065-1078.

Eaton, G.M., Brewer, G.J. and Tashian, R.E. (1966). Hexokinase isoenzyme patterns of human erythrocytes and leucocytes. *Nature 212*, 944-946.

Erhardt, F.W. and Scriba, P.C. (1977). High molecular thyrotropin ("Big"-TSH) from human pituitaries: preparation and partial characterization. *Acta endocr. 85*, 698-712.

Faloona, G.R. and Unger, R.H. (1974). Biological and immunological activity of pancreatic glucagon and enteric glucagon-like immunoreactivity. In: *Heterogeneity of Polypeptide Hormones*, pp. 142-149. Eds. Rabinowitz, D. and Roth, J. Academic Press, New York.

Freeman, C., Karoly, K. and Adelman, R.C. (1973). Impairments in the availability of insulin to liver *in vivo* and in the binding of insulin to purified hepatic plasma membrane during aging. *Biochem. Biophys. Res. Comm. 54*, 1573-1580.

Friesen, H., Guyda, H. and Hardy, J. (1970). Biosynthesis of human growth hormone and prolactin. *J. clin. Endocr. Metab. 31*, 611-624.

Gershon, H. and Gershon, D. (1973). Inactive enzyme molecules in aging mice: liver aldolase. *Proc. natn. Acad. Sci. USA 70*, 909-913.

Guidice, L.C. and Pierce, J.G. (1977). Separation of functional and nonfunctional β subunits of thyrotropin preparations by polyacrylamide gel electrophoresis. *Endocrinology 101,* 776-781.

Gold, G., Karoly, K., Freeman, C. and Adelman, R.C. (1976). A possible role for insulin in the altered capability for hepatic enzyme adaptation during aging. *Biochem. Biophys. Res. Comm. 73,* 1003-1010.

Golstein-Golaire, J. and Vanhaelst, L. (1975). Gel filtration profile of circulating immunoreactive thyrotropin and subunits of myxedematous sera. *J. clin. Endocr. Metab. 41,* 575-580.

Holland, J.J., Kohne, D. and Doyle, M.V. (1973). Analysis of viral replication in ageing human fibroblasts cultures. *Nature 245,* 316-318.

Jacob, F. and Monod, J. (1963). Genetic repression, allosteric inhibition, and cellular differentiation. In: *Cytodifferentiation and Macromolecular Synthesis,* pp. 30-64. Ed. Locke, M. Academic Press, New York.

Kemper, B., Habener, J.F., Mulligan, R.C., Potts, J.T. and Rich, A. (1974). Pre-proparathyroid hormone: a direct translation product of parathyroid messenger RNA. *Proc. natn. Acad. Sci. USA 71,* 3731-3735.

Klug, T.L. and Adelman, R.C. (1977). Evidence for a large thyrotropin and its accumulation during aging in rats. *Biochem. Biophys. Res. Comm. 77,* 1431-1437.

Koida, M., Lai, C.T. and Horecker, B.L. (1969). Subunit structure of rabbit muscle aldolase: extent of homology of α and β subunits and age-dependent change in their ratio. *Arch. Biochem. Biophys. 134,* 623-631.

Lee, G., Aloj, S.M., Beguinot, F. and Kohn, L.D. (1977). Existence of a soluble thyrotropin binding component in normal human sera. *J. Biol. Chem. 252,* 7967-7970.

Leuschen, M.P., Tobin, R.B., and Moriarity, C.M. (1978). Enriched populations of rat pituitary thyrotrophs in monolayer culture. *Endocrinology 102,* 509-518.

Lewis, U.J., Dunn, J.T., Bonewald, L.F., Seavey, B.K. and Vanderlaan, W.P. (1978). A naturally occurring structural variant of human growth hormone. *J. Biol. Chem. 253,* 2679-2687.

Martin, G.M. (1978). Genetic syndromes in man with potential revelance to the pathobiology of aging. In: *Genetic Effects on Aging,* pp. 5-40. Eds. Bergsma, D and Harrison, D.E. Alan R. Liss, Inc., New York.

Noe, B.D. and Bauer, G.E. (1971). Evidence for glucagon biosynthesis involving a protein intermediate in islets of the anglerfish *(Lophius americanus). Endocrinology 89,* 642-651.

Obenrader, M.F., Auth, J.C., Gold, G., Ceci, L., Kitahara, A. and Adelman, R.C. (1978). Heterogeneity of immunoreactive insulin in aging rats. *J. Geront.* (submitted).

Orgel, L.E. (1963). The maintenance of the accuracy of protein synthesis and its relevance to ageing. *Proc. natn. Acad. Sci.*

49, 517-521.
Papkoff, H., Sairam, M.R. and Li, C.H. (1971). Amino acid sequence of the subunits of ovine pituitary interstitial cell-stimulating hormone. *J. Am. chem. Soc. 93*, 1531-1532.
Peckham, W.D., Yamaji, T., Dierschke, D.J. and Knobil, E. (1973). Gonadal function and the biological and physicochemical properties of follicle-stimulating-hormone. *Endocrinology, 92*, 1660-1666.
Permutt, M.A. (1974). Effect of glucose on initiation and elongation rates in isolated rat pancreatic islets. *J. Biol. Chem. 249*, 2738-2742.
Permutt, M.A. and Routman A. (1977). Proinsulin precursors in isolated rat pancreatic islets. *Biochem. Biophys. Res. Comm. 78*, 855-862.
Permutt, M., Biesbroeck, J. and Chyn, R. (1977). Characteristics of high molecular weight insulins in insulinoma patients. *J. clin. Endocr. Metab. 44*, 536-544.
Pierce, J.G., Bahl, O.P., Cornell, J.S. and Swaminathan, N. (1971). Biologically active hormones prepared by recombination of the α chain of human chorionic gonadotropin and the hormone-specific chain of bovine thyrotropin or of bovine lutenizing hormone. *J. Biol. Chem. 246*, 2321-2324.
Pierce, J.G. (1974). Chemistry of thyroid-stimulating hormone. In: *Handbook of Physiological Chemistry*, pp. 79-103. Eds. Greep, R.O. and Astwood, E.B. Williams and Wilkins, Baltimore.
Prentice, L.G. and Ryan, R. (1975). LH and its subunits in human pituitary, serum and urine. *J. clin. Endocr. Metab. 40*, 303-312.
Rabinowitz, D. and Roth, J. (1974).(eds.) In: *Heterogeneity of Polypeptide Hormones*. Academic Press, New York.
Reichert, L.E., Jr. and Ramsey, R.B. (1977). Evidence for the existence of a large molecular weight protein in human pituitary tissue having follicle stimulating hormone activity. *J. clin. Endocr. Metab. 44*, 545-552.
Reiss, E. and Canterbury, J.M. (1974). Emerging concepts of the nature of circulating parathyroid hormones: implications for clinical research, In: *Recent Progress in Hormone Research*, pp. 391-429. Ed. Greep, R.O. Academic Press. New York.
Rinderknecht, E. Humbel, R.E. (1978). The amino acid sequence of human insulin-like growth factor I and its structural homology with proinsulin. *J. Biol. Chem. 253*, 2769-2776.
Roberts, J.L. and Herbert, E. (1978). Characterization of a common precursor to corticotropin and β-lipotropin. *Proc. natn. Acad. Sci. USA 74*, 5300-5304.
Roos, B.A., Okano, K. and Deftos, L.J. (1974). Evidence for a pro-calcitonin. *Biochem. Biophys. Res. Comm. 60*, 1134-1140.
Rosa, J. and Schapira, J. (1964). Lactic dehydrogenase isozymes and aging of erythrocytes. *Nature 204*, 883.
Rothstein, M. (1975). Aging and the alteration of enzymes: a review. *Mech. Aging and Dev. 4*, 325-338.

Rothstein, M. (1977). Recent developments in the age-related alterations of enzymes: a review. *Mech. Aging and Dev. 6*, 241-257.

Sherwood, L.M., Rodman, J.S. and Lundberg, W.B. (1970). Evidence for a precursor to circulating parathyroid hormone. *Proc. natn. Acad. Sci. USA 67*, 1631-1638.

Silverman, R. and Yalow, R.S. (1973). Heterogeneity of parathyroid hormone. *J. Clin. Invest. 52*, 1958-1971.

Skala-Rubinson, H., Vibert, M. and Dreyfus, J.C. (1976). Electrophoretic modifications of three enzymes in extracts of human and bovine lens: posttranslational "aging" of lens enzymes. *Clinica chim Acta 70*, 385-390.

Smith, L.F. (1964). Isolation of insulin from pancreatic extracts using carboxymethyl and diethylaminoethyl celluloses. *Biochim. biophys.Acta 82*, 231-236.

Smith, L.F. (1966). Species variation in the amino acid sequence of insulin. *Am. J. Med. 40*, 662-666.

Snyder, G., Hymer, W.C. and Snyder, J. (1977). Functional heterogeneity in somatotrophs isolated from rat anterior pituitary gland. *Endocrinology 101*, 788-799,

Steiner, D.F., Clark, J.L., Nolan, C., Rubenstein, A.H., Margoliash, E., Aten, B. and Oyer, P.E. (1969). Proinsulin and the biosynthesis of insulin. In: *Recent Progress in Hormone Research*, pp. 207-282. Ed. R.O. Greep. Academic Press, New York.

Steiner, D.F., Cunningham, D., Spigelman, L. and Aten, B. (1967). Insulin biosynthesis: evidence for a precursor. *Science 157*, 697-700.

Steiner, D.F., Holland, O., Rubenstein, A., Cho, S. and Bayliss, D. (1968). Isolation and properties of proinsulin, intermediate forms, and other minor components from crystalline bovine insulin. *Diabetes 17*, 725-736.

Strehler, B.L. (1977). *Time, Cells, and Aging*, pp. 295-324. Academic Press, New York.

Tompkins, G.A., Stanbridge, E.J. and Hayflick, L. (1974). Viral probes of aging in the human diploid cell strain WI-38. *Proc. Soc. exp. Biol. Med. 146*, 385-390.

Vaitukaitus, J.L. and Ross, G.T. (1974). Subunits of human glycoprotein hormones, their immunological and biological behavior. In: *Heterogeneity of Polypeptide Hormones*, pp. 98-106. Eds. Rabinowitz, D. and Roth, J. Academic Press, New York.

Vanhaelst, L. and Golstein-Golaire, J. (1976). Gel filtration profile of immunoreactive thyrotropin and subunits of human pituitaries. *J. clin. Endocr. Metab. 43*, 836-841.

Varandani, P.T. (1974). Insulin degradation VI. Feedback control by insulin of liver glutathione-insulin transhydrogenase in rat. *Diabetes 23*, 117-125.

Varandani, P.T., Shroyer, L.A. and Nafz, M.A. (1972). Sequential degradation of insulin by rat liver homogenates. *Proc. natn. Acad. Sci. USA 69*, 1681-1684.

Yalow, R. (1974). Heterogeneity of peptide hormones, In: *Recent Progress in Hormone Research* , pp. 597-633. Ed. Greep, R.O. Academic Press, New York.

Yalow, R. and Berson, S.A. (1970). Size and charge distinctions between endogenous human plasma gastrin in peripheral blood and heptadecapeptide gastrins. *Gastroenterology 58*, 609-615.

Yip, C.C., Hew, C.L. and Hsu, H. (1975). Translation of messenger ribonucleic acid from isolated pancreatic islets and human isulinomas. *Proc. natn. Acad. Sci. USA 72*, 4777-4779.

SOME NEUROENDOCRINE ASPECTS OF AGING

JAMES A. CLEMENS, RAY W. FULLER, and NORRIS V. OWEN

The Lilly Research Laboratories
Indianapolis, Indiana 46206

Many of the early studies on aging of the endocrine system were preoccupied with determining if senile changes in pituitary target organs were intrinsic to those organs or due to age-related changes in the pituitary gland. More recent studies suggest that the failure of endocrine organs may not be attributed to intrinsic changes in these organs or even senility of the pituitary gland, but rather to changes at some higher neural center that exerts a regulatory influence on pituitary tropin hormone secretion. Considerable emphasis is currently being placed on the hypothalamus as a center that may fail during the aging process, resulting in a cascading series of events that eventually lead to the senescence of endocrine organs as well as other parts of the body.

Here we will review some of the studies on aging of the neuroendocrine system as well as present some new data. We also will examine the rat as a model for aging of the neuroendocrine system and comment on the relevance of the rat model for the study of human aging.

REVIEW OF THE HYPOTHALAMUS-PITUITARY-GONADAL AXIS: RAT VERSUS HUMAN

Since the rat is being used as a model for human aging, a close examination of both differences and similarities between the two species is important. Similarities as well as important differences exist with respect to the hypothalamus-pituitary-gonadal (HGP) axis. An important question here appears to be the following one: Is there a common etiology for aging of the HPG axis in the rat and human? The above question can be restated in terms of two alternatives. Either the differences between rats and humans with respect to aging of the HPG axis result from some minor differences in hormone regulatory mechanisms, or the differences between the two species result from aging changes in entirely separate parts of the HPG axis. The usefulness of the rat as a model for human aging largely depends on which of the above alternatives can be supported by experimental evidence.

Many clinical investigators believe that in the human, menopause is brought on because the ovaries fail to respond to pituitary gonadotropins. The ovaries become involuted, are deficient in follicles and fail to secrete adequate amounts of steroids. Certainly in humans, there is no failure of gonadotropin production by the pituitary during aging; on the contrary loss of ovarian cycles during aging is characteristically followed by an increased output of luteinizing hormone (LH) and follicle stimulating hormone (FSH) *(Coble, Kohler, Cargille and Ross, 1969; Kohler, Ross and Odell, 1968; Tsai and Yen, 1971)*. The secretion of FSH increases about 15-fold after the menopause, and LH secretion increases about 3- to 10-fold. The high levels of circulating gonadotropins in postmenopausal women is a reflection of the low level of estrogens available to exert a negative feedback action on the hypothalamus. Since the negative feedback action of estrogen is gone, gonadotropin secretion continues unabated. In contrast to what happens to LH and FSH secretion in the human female, serum prolactin levels do not become elevated. Pepperell, Bright and Smith *(1977)* reported that serum prolactin levels in postmenopausal women were slightly lower than prolactin levels found in premenopausal women and in women taking oral contraceptive preparations (Table 1). Apparently in humans, serum prolactin is not elevated as a normal consequence of aging, but elevations are seen only in cases of pregnancy, galactorrhea or pituitary tumors.

TABLE 1. SERUM PROLACTIN LEVELS OF NORMAL WOMEN

Type of Subject	Number of Subjects	Serum Prolactin Levels (Mean ± SD) (ng/ml)
Premenopausal	78	10.6 ± 3.0
Postmenopausal	17	8.2 ± 4.0[a]
Taking combined oral contraceptive	33	12.2 ± 4.0[a]

[a] *$p<.01$*; [b] *$p<.025$*

From Pepperell et al., 1977.

The menstrual cycle in women is more irregular for some time following the menarche and for months and sometimes years preceeding the menopause than it is during the period of optimal reproductive capability *(Treolar, Boynton, Behn and Brown, 1967)*. Women approaching the menopause frequently have shorter cycles than they had earlier in their adult life, and the shorter cycles appear to be due to a reduction in the length of the follicular phase of the cycle *(Sherman and Korenman, 1975; Sherman, West and Korenman, 1976)*. After the menopause there is little evidence of reproductive cycles.

On the surface, aging of the HPG axis in the rat appears quite different from that in the human. In the rat, age-related changes in reproductive cycles follow a far less predictable pattern than has been observed in humans. The main reason for this difference is that the ovaries of rats continue to secrete steroids well after failure of the reproductive cycles.

Age-related changes in the reproductive cycles were characterized thoroughly by Ascheim *(1961)* and Bloch *(1961)*. Their work showed that following the cessation of normal reproductive cycles, rats have periods of constant estrus, repetitive pseudopregnancy, or both, which can last for many months. The ovaries in rats in constant estrus are small and contain a large number of follicles of variable size and no corpora lutea, while the pseudopregnant rat ovaries contain many large functional corpora lutea as well as follicles in various stages. In many laboratories the constant estrous state appears to be the most prevalent condition, and repetitive pseudopregnancies are the second most prevalent. Some rats demonstrate irregular estrous cycles followed by anestrus. Many animals enter into a condition of anestrus before death. A recent review by Ascheim *(1976)* gives an excellent description of the senile changes in the estrous cycles.

Secretion of gonadotropins in rats decreases with aging. The decrease in gonadotropin secretion is accompanied by an increase in prolactin secretion. Lower serum LH and FSH levels in old rats appear to result from an impairment of the release mechanisms for these hormones at the hypothalamic level. The elevation of serum prolactin levels in old rats may result from a combination of two factors. First, the inhibitory control of prolactin release by the hypothalamus might be reduced as a result of aging of the neuronal system responsible for releasing prolactin inhibiting factor. Second, the rat pituitary gland is unusually prone to development of prolactin secreting adenomata. In many old rats about 2 years of age large prolactin-secreting pituitary tumors are present. The tumors may have been allowed to form because of a progressively diminishing inhibitory influence from the hypothalamus and a constant stimulatory influence from estradiol. High serum levels of estrogens of several types are well known to induce pituitary tumors in rats.

Secretion of LH can be reinitiated in most aged female rats and usually leads to a resumption of ovarian function and estrous cycles. The constant estrous state has been studied most extensively in the rat and is characterized by the presence of well developed follicles in the ovary, absence of corpora lutea and absence of surges of LH in the blood. The absence of LH surges appears to be the cause of the constant estrous syndrome, because follicles continue to accumulate in the ovary and secrete estrogens. The failure to release LH probably results from age-induced changes in the brain, because stimulation of the preoptic area in senile constant estrous rats will induce ovulation and estrous cycles *(Clemens, Amenomori, Jenkins and Meites, 1969)*. In addition to preoptic area stimulation, several drugs and hormones as well as stress *(Linnoila, Markku and Cooper, 1976; Huang, Marshall and Meites, 1976)* have been shown to induce ovulation or reinstate vaginal cycles in old constant estrous rats (Table 2).

TABLE 2. EFFECT OF VARIOUS TREATMENTS ON CYCLES IN CONSTANT ESTROUS RATS

Treatment	Effect on Cycles
Epinephrine	Ovulation followed by a few normal cycles[a]
Preoptic stimulation	Ovulation followed by a few normal cycles[a]
L-DOPA	Many respond with normal cycles[b,c]
Progesterone	All show normal cycles and about 50% ovulate[a]
L-Tyrosine	Regular cycles[c]
ACTH	Regular cycles[b]
Ether Stress	Regular cycles[b]

[a] Clemens et al., 1969; [b] Huang et al., 1976; [c] Linnoila and Cooper, 1976.

Estrogen appears to exert a positive feedback effect on LH release in almost every mammalian species. In normal ovariectomized (OVX) adult rats estradiol administration induces a marked elevation of serum LH; however, in old rats estradiol appears to have no positive feedback effect of LH release. Peluso, Steger and Hafez

(1977) found that estradiol benzoate administration to 18-19 month-old female Wistar rats was unable to induce a surge of LH in the serum (Table 3).

TABLE 3. EFFECT OF ESTRADIOL BENZOATE ON SERUM LH IN AGED RATS

Age	No. of Rats	Treatment	Serum 24 hrs after	Serum 26 hrs after
Adult (5-6 Mos.)	3	Oil	28 ± 8	28 ± 8
	4	E.B.[b]	88 ± 24	91 ± 19
Old (18-19 Mos.)	3	Oil	32 ± 12	20 ± 8
	4	E.B.	29 ± 9	23 ± 4

a LH was expressed as ng NIAMD-LH-RP-1/ml serum
b E.B.=estradiol benzoate (0.5 μg/100 given S.C. for 5 days after ovariectomy followed by a single dose of 25 μg/100g)
Data from Peluso et al., 1977.

A similar type of experiment demonstrated that the ability of progesterone to release LH in estrogen-pretreated OVX rats was significantly reduced in 10- to 11-month old rats as compared to young 60- to 70-day old rats *(Lu, Huang, Chen, Kercz, Mioduszewsky and Meites, 1977)* (Fig. 1). In the above study, young cycling rats 60 to 70 days of age and irregularly-cycling or constant-estrous rats 10 to 11 months of age were ovariectomized. At 25 or 52 days after ovariectomy, they were each given a single subcutaneous injection of 80 μg/kg of estradiol benzoate, followed hours later by a single subcutaneous injection of 8 mg/kg of progesterone. Figure 1 shows that after progesterone administration LH rose significantly more in the young OVX rats than in the old OVX rats derived from irregular cyclers at 28 days after ovariectomy. At 55 days after ovariectomy, the LH rise after progesterone was attenuated in the group of OVX rats derived from old constant estrous rats and in the group of OVX rats derived from the irregular cycling rats.

The experiments of Pelluso *et al. (1977)* and Lu *et al. (1977)* serve to illustrate that with aging there is a loss of the positive feedback of steroids on LH release. This is very significant because the stimulatory influence of estrogen appears to be required

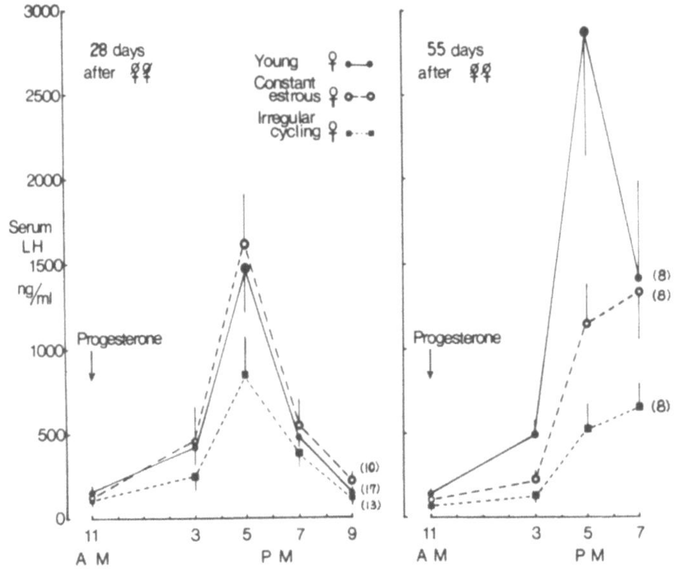

Figure 1. Serum concentration of LH 25 or 55 days after ovariectomy (☿☿) and treatment with a single dose of estrogen (EB) followed 3 days later with a dose of progesterone in young and old female rats. Prior to ovariectomy, the young rats showed regular estrous cycles and the old rats exhibited constant-estrus (CE) or irregular cycles (IRC). (n) = Number of rats per group. From Lu et al., 1977.

for spontaneous LH release in most mammalian species. The above observations in turn pose two interesting questions.

The first question is, how does ovulation occur in old pseudopregnant rats if the positive feedback of estrogen is gone? Perhaps progesterone plays a role in these rats because in the experiment by Lu *et al., (1977)* LH was released after progesterone injections, but the rats in the study wer not very old (10-11 months). It would be interesting to see the experiment repeated using animals about two years of age. The second question is, does estrogen exert its positive feedback effect via dopaminergic neurons? The stimulatory influence of L-dopa reported by Linoila, Markku and Cooper *(1976)* in old constant estrous rats appeared to be through activation of dopamine receptors. Perhaps estrogen exerts its positive feedback influence on a group of dopaminergic neurons which, during aging, die or lose their responsivity to estrogen. The feedback center for control of basal levels of LH by estrogen (negative feedback center) may not lose its sensitivity to estrogen, because basal serum levels of LH are not elevated.

Some similarities and differences in the HPG axis of rats and humans are summarized in Table 4. We made comparisons to point out that some basic differences exist in the aging HPG axis of rats and humans. At the present time, there is compelling evidence that aging changes in the CNS are a cause of reproductive senescence in the rat. In the human, little evidence exists to indicate that failure of the HPG axis results from aging changes in the brain. The possibility still exists, however, that aging changes in the CNS in the human may alter the pattern of gonadotropin secretion in such a way that ovaries undergo atrophic changes.

TABLE 4. SOME SIMILARITIES AND DIFFERENCES IN THE HPG AXIS OF RATS AND HUMANS

Observation	Rat	Human
Serum gonadotropins	Decrease	Increase
Serum prolactin	Increase	Decrease
Reproductive cycles	Abnormal	Stop
Ovarian function	Some functional capacity	Ovarian atrophy
Ovarian sensitivity to gonadotropins	Present	Absent
Positive estrogen feedback on LH	Absent	Absent
Negative estrogen feedback on LH	Present	Present
Histology of anterior pituitary	Many ademomata present	Active gonatrophs normal prolactin cells

RELATIONSHIP OF PROLACTIN TO SENILE CHANGES IN THE HPG AXIS

Elevated serum prolactin levels are inhibitory to gonadotropin secretion. The exact mechanism by which prolactin inhibits gonadotropin release is not known, but some propose a hypothalamic site of action, and others propose an action on the ovarian follicle. In humans, prolactin has been suggested to induce hypogonadism by a direct action on the ovary *(Thorner, McNeilly, Hagan and Besser, 1974)*. They propose that the hypogonadism in patients with pathological lactation is secondary to an inappropriately raised prolactin level, since the hypogonadism as well as the galactorrhea respond to bromocryptine, which specifically blocks prolactin secretion. The reduction in prolactin appears to allow the circulating gonadotropins to act on the gonad to stimulate steroidogenesis. The sex steroids may feed back on the hypothalamus and pituitary to allow normal release to gonadotropins, leading to a return of normal gonadal function. A study by Rolland, DeJong, Schellekens and Lequin *(1975)* tends to support the above hypothesis.

Other human studies suggest that prolactin might inhibit gonadotropin secretion by an action on the CNS or possibly on the pituitary *(del Pozo, Varga, Schulz, Kunziz, Marbach, del Campo and Eppenberger, 1975; Pepperell, Evans, Brown, Bright, Smith, Burger and Healy, 1977)*. In addition, the report that stimulation of human prolactin release with TRH would suppress the pulsatile pattern of LH secretion suggests an antigonadotropic action of prolactin at the pituitary or hypothalamic level *(Bohnet, Dahlen and Schneider, 1974)*. A defect in hypothalamic-pituitary function is again suggested by the fact that most hyperprolactinemic patients are relatively hypoestrogenic, and serum gonadotropin levels are not elevated. Further supporting a central nervous system defect is the fact that normal short-term fluctuations of LH concentrations are lost in such patients *(Bohnet, Dahlen, Wuttke and Schneider, 1976)*. That the hypothalamus rather than the pituitary may be primarily responsible is suggested by several studies showing a normal gonadotropin response to infused LHRH in most hyperprolactinemic patients, particularly those without tumors. Thus, prolactin may act at more than one locus in the HPG axis to impair gonadotropic function. Wherever the site of action, however, it is abundantly clear that high serum prolactin levels block cyclic release of LH.

In the rat, evidence exists for an inhibitory influence of high serum levels of prolactin on LH release. Injections of LHRH produced significantly less LH release in lactating rats than in normal cycling female rats on diestrous day 2, and the pituitaries from lactating rats released less LH *in vitro* than in the pituitaries from cyclic rats released *(Lu, Chen, Grandison, Huang, and Meites, 1976)*. Grandison, Advis, Hodson, Simpkins and Meites *(1976)* used pituitary isografts to increase serum prolactin levels and found that in female rats the rise in serum LH was significantly reduced by prolactin at 5 and 9 days after ovariectomy. Beck, Engelbart, Gelato and Wuttke *(1977)* reported that elevated serum prolactin levels, produced by means of pituitary isografts, abolished pulsatile LH release in OVX rats. In the study of Beck *et al. (1977)*, LH suppression only lasted from 4-6 days after the pituitary isografts were placed, but in the study of Grandison *et al. (1976)*, LH suppression was still strong 9 days after placement of the pituitary grafts.

The aforementioned inhibitory properties of prolactin on gonadotropin secretion may help explain some of the differences between the human and the rat with regard to aging of the HPG axis. The inhibitory influence of prolactin on LH release is not present in old humans, because serum prolactin levels are very low *(Pepperell et al., 1977)*. However, in the old rat perhaps the low serum LH levels are due to the extremely high serum prolactin levels. The possibility exists that old rats might regain a normal capacity to release LH if the serum prolactin levels were reduced. We decided to investigate this possibility using the dopamine agonist, lergotrile mesylate, to lower serum prolactin levels.

In the first study, 28-month-old Wistar rats were compared to young 3-month-old rats in one experiment and to 9-month-old Wistar rats in the second experiment. An attempt was made to select animals in good health and free from any gross pathology. Most of the old rats were in a state of continuous diestrus; however, during the course of treatment some proestrous and estrous vaginal smears were noted. The young rats and the old rats were treated for 60 days with 1.5 mg/kg of lergotrile mesylate injected intraperitoneally daily. Corresponding control groups were treated daily for 60 days with intraperitoneal injections of 0.2 ml of sterile water, which was the vehicle for lergotrile mesylate. At the end of the treatment period, the rats were selected on a day of diestrus and were anesthetized with metaphane, and a 1.0 ml blood sample was collected by jugular puncture. Serum prolactin was assayed using the NIAMDD radioimmunoassay kit and serum LH was assayed by radioimmunoassay according to the method of Niswender, Midgley, Monroe and Reichert *(1968)*.

Table 5 shows that chronic suppression of prolactin for a two-month period by lergotrile mesylate did not produce an elevation of serum LH levels in experiment 1 or in experiment 2. In fact, in experiment 1 lergotrile mesylate treatment resulted in a significant reduction of serum LH in the 30-month-old rats, but not in the young rats. Interestingly, by 11 months serum prolactin levels were already elevated considerably when compared with the levels from 5-month-old rats. The results of these experiments indicate that the high serum prolactin levels in old rats that no longer demonstrate reproductive cycles do not suppress basal serum levels of LH, since no "postmenopausal-like" rise of LH was observed after inhibition of prolactin secretion. Although basal levels of LH were not elevated, infrequent surges of LH may have been induced by the lowered serum prolactin levels, because the predominantly diestrous smear pattern was interrupted occasionally in some rat by a day or two of estrus. This observation is consistent, however, with the reports in humans that elevated serum prolactin levels prevent cyclic LH release.

Another of the apparent deficits in gonadotropin secretion in old rats that we had previously mentioned was the attenuation of the post-ovariectomy rise of serum LH. Since Grandison *et al. (1976)* reported that high serum prolactin levels in young animals could blunt the post-ovariectomy rise in LH, we decided to determine if inhibition of prolactin secretion in 2-year-old OVX female

TABLE 5. EFFECT OF LERGOTRILE MESYLATE ON SERUM PROLACTIN AND LH LEVELS IN YOUNG AND OLD RATS

Group and Treatment	No. of Rats	Serum Prolactin (ng/ml)	Serum LH (ng/ml)
	EXPERIMENT I		
30-month-old rats, water injection	4	181.7 ± 7.2[a]	15.6 ± 3.7
30-month-old rats, lergotrile mesylate (1.5 mg/kg)	5	9.4 ± 3. ($p<.001$)	3.1 ± 2.0 ($p<.05$)
5-month-old rats, water injection	6	19.1 ± 5.2	23.9 ± 6.4
5-month-old rats, lergotrile mesylate	6	6.6 ± 5.2	13.2 ± 1.8
	EXPERIMENT II		
30-month-old rats, water injection	9	206.5 ± 18.8	18.8 ± 1.0
30-month-old rats, lergotrile mesylate (1.5 mg/kg)	10	38.5 ± 12.6	18.7 ± 3.0
11-month-old rats, water injection	6	155.4 ± 20.8	22.6 ± 4.2
11-month-old rats, lergotrile mesylate (1.5 mg/kg)	6	8.0 ± 2.4 ($p<.001$)	29.5 ± 5.5

[a] *Mean ± SE*

Wistar rats would allow serum LH levels to rise to levels that are comparable with LH levels obtained after ovariectomy in young 6-month-old rats.

In this study 2-year-old pseudopregnant and normally cycling Wistar rats were OVX. The day after ovariectomy they were divided into four groups. Group 1 contained 7 OVX 2-year-old rats that received injections of sterile water. Group 2 contained 7 OVX 2-year-old rats that received injections of 3.0 mg/kg of lergotrile mesylate. Group 3 was made up of 6 OVX 6-month-old rats that received sterile water injections, while group 4 contained 6 similar rats that received 3.0 mg/kg of lergotrile mesylate. All injections were given daily and intraperitoneally and the treatment lasted 28 days. At the end of the treatment period the rats were decapitated, and blood was collected for assay of prolactin and LH.

Table 6 shows that inhibition of prolactin secretion did not result in a larger post-ovariectomy rise in serum LH. In fact, in these 2-year-old Wistar rats no post-ovariectomy rise in LH was observed in either of the two old groups. This becomes evident when referring back to Table 5 to the LH levels found in intact old female rats. In the present experiment treatment with lergotrile mesylate appeared to reduce serum LH levels in the old OVX rats, while in the young rats it appeared to increase serum LH levels. The reason for this difference is not clear, but in most of our aging studies lergotrile mesylate appears to potentiate LH release in young adults and either not effect or decrease basal LH levels in old rats.

TABLE 6. EFFECT OF LERGOTRILE MESYLATE ON LH AND PROLACTIN SECRETION IN OVX YOUNG AND OLD RATS

Group and Treatment	No. of Rats	Serum Hormone Levels Prolactin (ng/ml)	Serum Hormone Levels LH (ng/ml)
1. Old rats, sterile water	7	191.8 ± 23.4[a]	28.7 ± 7.6
2. Old rats, lergotrile mesylate (3.0 mg/kg)	7	26.2 ± 14.7 (p<0.1)[b]	10.7 ± 3.7
3. Young rats, sterile water	6	6.9 ± 1.62 (p<.01)[c]	269.1 ± 49.4 (p<.001)[c]
4. Young rats, lergotrile mesylate (3.0 mg/kg)	6	1.7 ± 0.2 (p<.01)[b,c]	391.7 ± 60.7 (p<.001)[c]

[a] *Mean ± SE*
[b] *When compared with its respective control*
[c] *When compared with group 1*

In the next study, we investigated the influence of lergotrile mesylate on the estrogen-induced surge of LH. We reasoned that if prolactin is inhibitory to cyclic LH release, inhibition of prolactin secretion might increase the amount of LH released in the old OVX rat in response to estradiol treatment. Twenty 4-month-old rats and 3-month-old rats were ovariectomized and received no further treatment for 1 month. After 1 month all rats received subcutaneous injections of 3 μg of estradiol benzoate in oil daily for 6 days. Half the rats also received daily intraperitoneal injections of 3.0 mg/kg of lergotrile mesylate for 6 days. At 0800 hours and at 1600 hours on the 7th day the rats were anesthesized with metaphane and blood was collected for the determination of prolactin and LH by radioimmunoassay.

The results are shown in Table 7. Estradiol benzoate administration produced an elevation of serum LH levels in the old OVX rats. Every old rat showed an afternoon elevation (group 1), but the mean was not significantly higher using the "paired t-test" ($p=0.07$) because of variation in individual values; however, when the non-parametric sign test was used the afternoon increase became highly significant($p<.01$). Afternoon serum LH levels in lergotrile mesylate-treated rats were not significantly elevated. In the young rats, significant afternoon surges of serum LH were seen in rats that received estradiol benzoate and in rats that received estradiol benzoate and lergotrile mesylate. The mean serum LH level at 1600 hours in young rats receiving lergotrile mesylate was higher than that in the rats receiving estradiol benzoate alone. The results of the above studies suggest that the inability of rats about 24-26 months of age to show a positive LH feedback response to estradiol benzoate is not due to high serum prolactin levels. The reason why these rats are unable to generate LH surges in response to an estrogenic stimulus is not completely clear at this time but almost certainly is associated with aging of some CNS area. Tsai and Yen *(1971)* reported that in postmenopausal humans the positive feedback effect of estrogens on LH secretion was absent. In this respect, the rat and human show a similar age-related change. With respect to basal levels of LH, the rat and human are quite different. The inability of the old rat to show elevated serum LH levels after inhibition of prolactin secretion and after ovariectomy plus inhibition of prolactin secretion indicates that basic differences exist between humans and rats in the aging of the mechanisms that regulate basal LH secretion.

Recently, Clemens and Bennett *(1977)* reported that vaginal cyclicity could be reinstated by daily injections of lergotrile mesylate in approximately 2-year-old pseudopregnant female Wistar rats. They also found that preoptic area lesions in young rats induced repeated periods of pseudopregnancy and that lergotrile mesylate administration to the young rats with lesions reinstated vaginal cyclicity (Table 8). The ovaries of the old pseudopregnant rats contained corpora lutea, thus LH release and ovulation should have occured. Surges of LH and ovulation in the old pseudopregnant

TABLE 7. EFFECT OF LERGOTRILE MESYLATE ON THE ESTRADIOL-INDUCED SURGE OF LH IN OLD AND YOUNG RATS

Group and Treatment	No. of Rats	Serum LH levels (ng/ml) 0900 hr	1600 hr
1. Old rats, 3 μg E.B.[a]	7	25.1 ± 5.3[b]	41.8 ± 11.5[c]
2. Old rats, 3 μg E.B. + lergotrile mesylate (3 mg/kg)	7	22.6 ± 4.7	29.1 ± 8.8
3. Young rats, 3 μg E.B.	6	126.8 ± 25.6	586.6 ± 91.6 ($p<.01$)
4. Young rats, 3 μg E.B. + lergotrile mesylate (3 mg/kg)	6	98.1 ± 12.4	1191.4 ± 252.8 ($p<.01$)

[a] *E.B. = Estradiol benzoate;* [b] *Mean ± Standard error*
[c] *p = .07 using "paired t-test", and p<.01 using sign test (1600 vs 0900 hr)*

rats appear, at least on the surface, to be incompatible with what we just reported: that is, greatly attenuated positive feedback of estrogen on LH in old pseudopregnant rats. We decided to examine old pseudopregnant rats for LH surges on the afternoon of proestrus, because of the diminished positive feedback of estradiol on LH release. Wistar rats, 23-25 months of age and 3 months of age, were used in this experiment. Daily vaginal smears were recorded and rats showing repeated periods of psuedopregnancy were selected for experimentation. Only rats free from infection, tumors or other pathology were used in this study. The rats were housed in a room with a 14-hr light and 10-hr dark cycle (lights on at 0400 hr). On the day of proestrus, as indicated by vaginal smears, 10 old rats and 10 young rats were decapitated at 1530 hours, and the blood was collected and assayed for LH. In the old rat group, 5 of the rats showed proestrous LH surges. The remaining 5 showed no evidence of surges. Table 9 shows that the LH surge in the old rats was as high as the surge seen in young animals. This finding was surprising in view of all of the evidence showing that old rats were deficient in their ability to release LH under a variety of circumstances.

TABLE 8. EFFECTS OF LERGOTRILE MESYLATE ON ESTROUS CYCLES IN OLD PSEUDOPREGNANT RATS AND IN YOUNG RATS WITH MEDIAL PREOPTIC AREA LESIONS [a]

Group	No. of Rats	Normal 4-or5-day cycles	Longer 6-or7-day cycles	Pseudo-pregnancies
Old pseudo-pregnant	13	9	4	0
Young rats with medial preoptic area lesions	20[b]	17	0	0

[a] *The daily dose of lergotrile mesylate was 4.5 mg/kg.*
[b] *Three rats went into constant estrus during lergotrile mesylate treatment.*

TABLE 9. SERUM LH LEVELS ON THE AFTERNOON OF PROESTRUS IN OLD AND YOUNG RATS

Group	No. of Rats	Serum LH Levels (ng/ml)
23-to 25-month old rats	5	551 ± 45[a]
3-month old rats	10	604 ± 86

[a] *Mean ± SE*

The fact that old pseudopregnant rats do show LH surges but only show small surges of LH in response to estrogen indicates that some factor in addition to estrogen is necessary in the old rat to induce the LH surge. We are assuming here, however, that the estrogen-induced surge of LH in ovariectomized rats accurately reflects the physiological events that occur in intact animals. This assumption may be incorrect. Estradiol treatment followed by progesterone administration to ovariectomized old rats did induce a small LH surge *(Lu et al., 1977)*.

The failure to obtain surges of LH in the remaining 5 rats could be due to several factors. First, the day chosen to kill the rats may not have been the day of proestrus, because occasionally a day of proestrus and two days of estrus are seen between pseudopregnancies. Second, perhaps the surges of LH in some of the old rats occur later in the day. Later and diminished LH surges have been reported to occur in old rats *(van der Shoot, 1976)*. Third, maybe no LH surge occurred at all and the cycle was anovulatory. The possibility exists that several of the normally appearing cycles induced by lergotrile mesylate in old pseudopregnant rats might be anovulatory.

One item that becomes apparent from these studies is that the mechanism controlling basal levels of LH secretion and the mechanism than controls the cyclic surge of LH are clearly separate entities. The neural mechanism regulating basal levels of LH is deficient in rats because little or no elevation of serum LH occurs in the presence of diminished ovarian activity or after ovariectomy, while in the human, this neural mechanism is not compensated because humans have elevated serum LH levels after ovarian regression. The neural mechanism regulating the cyclic release of LH may age in a similar fashion because in both humans and rats the cyclic LH release mechanism does not appear to function properly.

DOPAMINERGIC FUNCTION AND AGING

One of the neurotransmitter systems that demonstrates age-related changes is the dopaminergic system. In humans, Parkinson's disease appears, at least in part, to result from an age-related dysfunction of the nigro-neostriatal (A9) dopaminergic system. Since many different dopaminergic systems are found in the brain, several different aspects of aging might be influenced by alterations of dopaminergic function, providing of course, that the various dopaminergic systems deteriorate with age. Age-related changes in dopaminergic function may play a role in the dysfunction of the neuroendocrine system that occurs with old age.

Linnoila, Markku and Cooper *(1976)* reported that L-dopa reinstates vaginal cycling in old constant estrous rats by stimulating dopamine receptors. They found that the effectiveness of L-dopa was potentiated by a peripheral decarboxylase inhibitor, blocked by pimozide, and unaffected by phenoxybenzamine or propranolol. Although L-dopa appears to reinstate vaginal cyclicity in old constant estrous rats, many seemingly unrelated treatments also appear to reinitiate vaginal cycles and ovulation (see Table 2). At the present time it is not possible to say whether any or all of the treatments in Table 2 eventually result in activation of dopamine receptors.

Repeated periods of pseudopregnancy can be induced in young adult rats by placing lesions in the medial preoptic area. The pseudopregnancies in these rats resemble the repeated pseudopregnancies that occur with aging. Regular cycles can be reinstated in both the lesioned rats and in the old rats by lergotrile mesylate administration *(Clemens and Bennett, 1977)*. In addition to inhibition of prolactin secretion, lergotrile mesylate may have restored cyclic endocrine function by stimulation of brain dopamine receptors. Dopamine receptor stimulation appears to be the mechanism by which lergotrile mesylate inhibits prolactin secretion *(Clemens, Smalstig and Shaar, 1975)*. The preoptic area lesions may have destroyed dopaminergic neurons that pass through or originate in the lesioned area and project to some locus where they participate in controlling cyclic regularity. After dopamine neurons degenerate, the dopamine receptors on the postsynaptic cells remain and become supersensitive *(Ungerstedt, Ljungberg, Hoffer and Siggins, 1975)*. Thus, the denervated dopamine receptors on neurons that control cyclic regularity may have been stimulated by lergotrile mesylate. Perhaps in old pseudopregnant rats a population of dopamine neurons may have degenerated or may have begun functioning in an abnormal fashion as a result of the aging process. Lergotrile mesylate administration merely restored the necessary dopaminergic stimulus needed for cycling to occur.

The similarities between young adult rats with preoptic area lesions and old pseudopregnant rats suggest that cyclic abnormalities in aged animals may be triggered by changes in the CNS. Repeated periods of pseudopregnancy in old rats may be due to some age-related change in the CNS. This view is supported by studies reporting abnormalities in catecholamine metabolism in the brains of rats. Table 10 summarized some of the changes in dopaminergic systems reported to occur with advancing age. Miller, Shaar and Riegle *(1976)* reported decreased hypothalamic dopamine levels in old male rats, and Simpkins, Mueller, Huang and Meites *(1977)* reported decreased dopamine levels as well as decreased hypothalamic dopamine turnover in old male rats. Both groups also reported a decrease in norepinephrine levels in the old male rats.

In view of the importance of catecholamines in cyclic endocrine function and the observation that the dopamine agonist, lergotrile mesylate, reinstated vaginal cyclicity in old pseudopregnant rats, we decided to measure levels of dopamine, norepinephrine and epinephrine in hypothalami in old female rats that had been treated with lergotrile mesylate or a water vehicle.

Two-year-old pseudopregnant and 6 month old normal cycling female Wistar rats were used in this study. One group of old pseudopregnant rats received intraperitoneal injections of 1.5

mg/kg of lergotrile mesylate daily for 90 days, while another group of old pseudopregnant rats (control group) received daily intraperitoneal injections of sterile water (0.2 ml). The 6-month-old rats received similar treatments. At the end of the treatment period the old rats were 27 months of age, and the cycling adults were 9 months of age. Daily vaginal smears were recorded during the treatment period, and 24 hours after the last injection the rats were decapitated during a day of diestrus. Brains were immediately removed and an approximately 25 mg fragment of basal hypothalamic tissue was removed and quickly frozen on Dry Ice. Hypothalamic concentrations of dopamine, norepinephrine and epinephrine was determined using high pressure liquid chromatography coupled with electrochemical detection according to the method of Fuller and Perry *(1978)*.

TABLE 10. CHANGES IN DOPAMINERGIC SYSTEMS REPORTED TO OCCUR WITH ADVANCING AGE

Area of Brain	Observation	References
Substantia nigra (human)	decreased tyrosine hydroxylase	Cote and Kremzner (1974)
Neostriatum (rat)	decreased tyrosine hydroxylase	McGeer *et al.*, (1971)
Whole brain minus neostriatum and cerebellum (rat)	no change in dopamine level	McGeer *et al.*, (1971)
Hypothalamus (rat)	decreased dopamine level	Miller *et al.*, (1976)
Hypothalamus (rat)	decreased dopamine level and turnover	Simpkins *et al.*, (1977)
Olfactory tubercle and caudate nucleus (rat)	decreased tyrosine hydroxylase	Reis *et al.*, (1977)

In the 27-month-old rats hypothalamic dopamine and epinephrine concentrations were significantly reduced when compared to the 9-month-old control values (Table 11). Lergotrile mesylate treatment in the aged rats restored hypothalamic dopamine levels to levels found in the 9-month-old controls. Lergotrile mesylate treatment did not alter hypothalamic norepinephrine or epinephrine in any group. In this experiment no change was found in hypothalamic norepinephrine. Other studies *(Miller et al., 1976); Simpkins et al., 1977)* reported a decrease in hypothalamic norepinephrine in old male rats.

The results of this preliminary study indicate that an age-related reduction in dopamine and epinephrine concentration occurs in the hypothalamus of old pseudopregnant rats and that the reduction in dopamine can be counteracted by lergotrile mesylate. The reduction of hypothalamic dopamine concentration adds support for the hypothesis that disruption of cyclic endocrine function in old female rats is related to some CNS abnormality in dopamine metabolism. Several possible explanations exist for the increase in hypothalamic dopamine concentration in the old rats and this topic is receiving further study. Very little is known about the role of epinephrine in the brain, especially in aging animals. This is the first report of the measurement of brain epinephrine in aged rats, and more studies will be needed to interpret the meaning of the decrease we observed.

TABLE 11. EFFECTS OF LERGOTRILE MESYLATE ON HYPOTHALAMIC CONCENTRATION OF DOPAMINE, NOREPINEPHRINE AND EPINEPHRINE IN 27-MONTH-OLD AND 9-MONTH-OLD FEMALE RATS

Group and Treatment	No. of Rats	Hypothalamic Concentration (pmoles/mg) Dopamine	Norepinephrine	Epinephrine
27-month-old (control)	6	2.3 ± 0.29[a]	13.0 ± 1.0	0.23 ± 0.03[a]
27-month-old (1.5 mg/kg lergotrile mesylate)	6	3.2 ± 1.6[b]	15.1 ± 1.6	0.29 ± 0.05
9-month-old (control)	5	3.5 ± 0.40[b]	15.4 ± 1.0	0.38 ± 0.06[b]
9-month-old (1.5 mg/kg lergotrile mesylate)	4	3.3 ± 0.12[b]	14.5 ± 0.8	0.37 ± 0.05[b]

a,b Values in each column with different superscripts are significantly different from each other (p<.05). Means ± SE

USE OF DOPAMINE AGONISTS TO TREAT SYMPTOMS OF AGING

At the moment, our scientific knowledge is not sufficient to enable us to make a rational judgement regarding the value of treating the symptoms of senility. At best, the agents currently available for treating disturbances of mental function in the elderly are only mildly effective. Research on the mechanism of action of drugs presently showing some beneficial effects appears to be one of the best approaches toward the eventual development of therapeutic agents in this area.

One drug that appears to possess some limited value in geriatric medicine is Hydergine®. This drug is a mixture of dihydroergocornine, dihydroergocryptine and dihydroergocrystine. A recent review by Hughes, Williams and Currier (36) indicated that Hydergine® consistently produced statistically significant improvement in several symptoms associated with dementia; however, there was only a small degree of improvement.

Hydergine® is thought to exert its beneficial effect on brain function in the elderly by improving blood flow to the brain or by stimulating the activity of energy-producing enzymes. Hydergine® also had been used to treat hypertension because of its alpha-blocking properties.

One property of many ergolines, the ability to stimulate dopamine receptors *(Corrodi, Fuxe, Hokfelt, Lidbrink and Ugerstedt, 1973)*, had apparently not been considered as an explanation of the Hydergine® effect. We found that ergoline derivatives inhibit prolactin release by acting on dopamine receptors located in the pituitary gland *(Clemens et al., 1975)*. We decided to determine if Hydergine® had any prolactin-inhibiting properties. Adult male rats were primed with 2.0 mg of reserpine, intraperitoneally and received an intraperitoneal injection of 40 μg/kg of Hydrogine®, bromocryptine or lergotrile mesylate 24 hours later. Rats were killed by decapitation 1 hour after treatment with the ergolines, and the blood was assayed for prolactin by redioimmunoassay. Table 12 shows that Hydergine® is a very potent inhibitor of prolactin secretion. The inhibition by Hydergine® is approximately equivalent to that shown by the two known dopamine agonists, bromocyrptine and lergotrile mesylate. These results indicate that, in addition to cerebral vascular effects, Hydergine® may act in some areas of the brain as a dopamine agonist, similar to the activity mentioned previously for lergotrile mesylate. Perhaps the central dopaminergic activity of Hydergine® is responsible for its beneficial effects in treating organic brain dysfunction. Another interesting aspect of the pharmacology of dopaminergic ergolines is the effect of lergotrile mesylate of longevity. When lergotrile mesylate was fed daily to rats during a two-year period there was a significant increase in the lifespan of female rats in two separate studies. Male and female Wistar rats were fed diets containing 0.0 (60 rats), 0.0015 (40 rats), 0.0045 (40 rats), 0.01 (40 rats), percent lergotrile mesylate. The studies were performed in duplicate.

TABLE 12. EFFECTS OF VARIOUS ERGOT ALKALOIDS ON SERUM PROLACTIN LEVELS IN RESERPINIZED MALE RATS

Group	No. of Rats	Serum Prolactin Levels (ng/ml)
Control (10% ethanol-saline vehicle)	10	20.3 ± 2.1[a]
Hydergine® (40 μg/kg)	10	8.5 ± 1.1 ($p<.01$)
Bromocryptine (40 μg/kg)	10	7.1 ± 0.9 ($p<.01$)
Lergotrile mesylate	10	7.0 ± 0.9 ($p<.01$)

[a] *Mean ± SE*

The female rats from both studies that were treated with lergotrile mesylate tended to live longer than the controls (Figures 2 and 3). Lergotrile mesylate had no consistent effect on the survival of males. All animals treated with lergotrile mesylate consumed the same amount of food; however, the treated rats (both males and females) tended to have somewhat lower body weights than the controls. The increased longevity of the females does not appear to be due to decreased food consumption or reduction in body weight, because the males showed the same small reduction in body weight.

Recently, Cotzias, Miller, Tang, Papavasiliou and Wang *(1977)* reported that administration of L-dopa to mice in their diet prolonged the mean-life span by a maximum of 50 percent. The findings with L-dopa are similar to our findings reported here with lergotrile mesylate. At the present time the precise reason why dopaminergic compounds increase longevity is not clear, but the above preliminary findings favor a more intensive study of the effects of dopaminergic agents on the aging process.

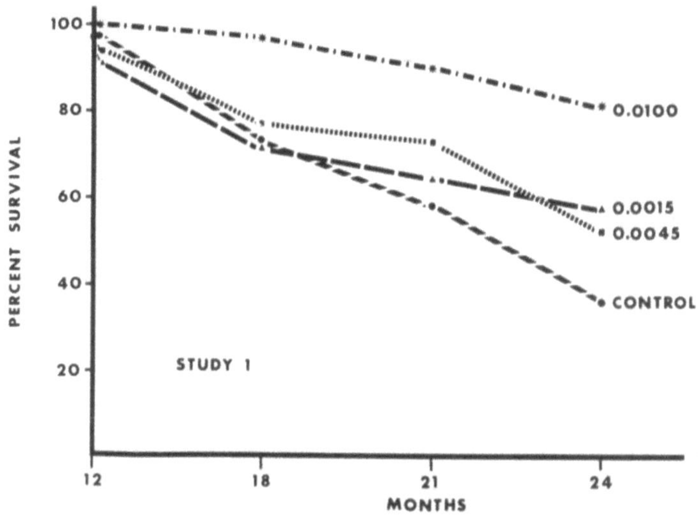

Figure 2. Survival curves for rats fed various daily doses of lergotrile mesylate. The percent (by weight) of the diet which was lergotrile mesylate in indicated on the figure.

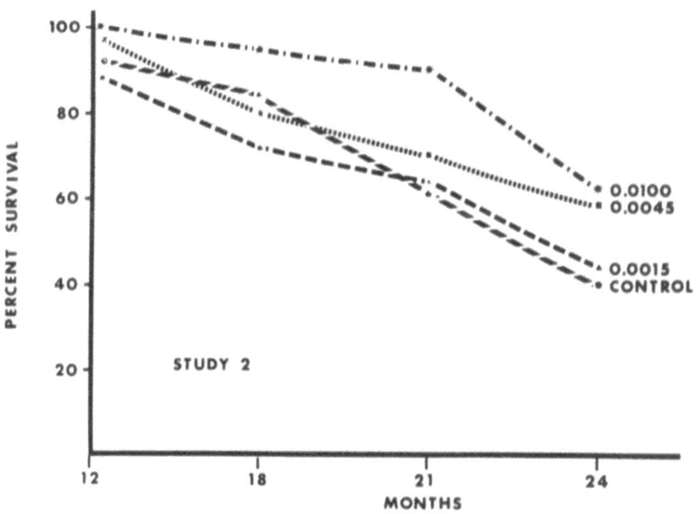

Figure 3. Survival curves for rats fed lergotrile mesylate in the diet. See Figure 2.

ACKNOWLEDGEMENTS

We thank the NIAMDD Rat Pituitary Hormone distribution program, Dr. Gordon Niswender and Dr. L. E. Reichert, Jr. for radioimmunoassay materials. We also wish to thank Mr. David Bennett and Mr. Kenneth Perry for technical assistance.

REFERENCES

Ascheim, P. (1961). La pseudogestation a repetition ches les rattes seniles. *C.R. Acad. Sci. (Paris) 253:* 1988-1933.
Ascheim, P. (1976). Aging in the hypothalamic-hypophyseal ovarian axis in the rat. In: *Hypothalamus Pituitary and Aging.* pp. 376-418. Eds. A.V. Everitt, and J.A. Burgess, Charles C. Thomas, *Springfield.*
Beck, W., Engelbart, S., Gelato, M. and Wuttke, W. (1977). Antigonadotrophic effect of prolactin in adult castrated and in immature female rats. *Acta endocr. 84,* 62-71.
Bohnet, H.G., Dahlen, H.G., Schneider, H.P.G. (1974). Hyperprolactinemia and pulsatile LH fluctuation. *Acta endocr. (Supp) 184,* 109.
Bohnet, H.G., Dahlen, H.G. Wuttke, W. and Schneider, H.P.G. (1976). Hyperprolactinemic anovulatory syndrome. *J. clin. Endocr. Metab. 42,* 132-143.
Bloch, S. (1961). Investigations of genital aging of female rats. *Gerontologia 5,* 55-62.
Clemens, J.A., Amenomori, Y., Jenkins, T. and Meites, J. (1969). Effects of hypothalamic stimulation, hormones, and drugs on ovarian function in old female rats. *Proc. Soc. exp. Biol. Med. 132,* 561-563.
Clemens, J.A. and Bennett, D.R. (1977). Do aging changes in the preoptic area contribute to loss of cyclic endocrine function? *J. Geront. 32,* 19-24.
Clemens, J.A., Smalstig, E.B. and Shaar, C.J. (1975). Inhibition of prolactin release by lergotrile mesylate: mechanism of action. *Acta endocr. 79,*230-237.
Coble, Y.D., Jr., Kohler, P.O., Cargille, C.M. and Ross, G.T. (1969). Production rates and metabolic clearance rates of human FSH in pre-menopausal and post-menopausal women. *J. clin. Invest. 48,* 359-363.
Corrodi, H., Fuxe, K.,Hokfelt, T., Lidbrink, P., and Ungerstedt, U. (1973). Effect of ergot drugs on central catecholamine neurons: evidence for a stimulation of central dopamine neurons. *J. Pharm.Pharmac. 25,* 409-412.
Cote, L.J. and Kremzner, L.T. (1974). Changes in neurotransmitter systems with increasing age in human brain, In: *Transactions of the American Society for Neurochemistry.* 5th Annual Meeting, p. 83.

Cotzias, G., Miller, S.T., Tang, L.C., Papavasiliou, P.S. and Wang, Y.Y. (1977). Levodopa, fertility and longevity. *Science 196*, 549-551.

del Pozo, E., Varga, L., Schulz, K.D., Kunziz, H.J., Marbach, P., del Campo, G.L. and Eppenberger, U. (1975). Pituitary and ovarian response patterns to stimulation in the postpartum and in galactorrhea-amenorrhea: The role of prolactin. *Ob. Gyn. 46*, 539-543.

Fuller, R.W. and Perry, K.W. (1978). *Res. Comm. Chem. Path. Pharmac.* (in press).

Grandison, L., Advis, J., Hodson, C., Simpkins, J. and Meites, J. (1976). Effects of prolactin on postcastration LH release. *ICRS Med. Sci. 4*, 427.

Huang, H.H., Marshal, S. and Meites, J. (1976). Induction of estrous cycles in old non-cyclic rats by progesterone, ACTH, ether stress or L-DOPA. *Neuroendocrinology 20*, 21-34.

Hughes, J.R., Williams, J.G. and Currier, R. (1976). An ergot alkaloid preparation (Hydergine®) in the treatment of dementia: Critical review of the literature. *J. Am. Geriat. Soc. 24*, 490-497.

Kohler, P.O., Ross, G.T. and Odell, W.D. (1968). Metabolic clearance and production rates of human luteinizing hormone in pre- and post-menopausal women. *J. clin. Invest. 47*, 38-47.

Linnoila, Markku, and Cooper, R.L. (1976). Reinstatement of vaginal cycles in aged female rats. *J. Pharmac. exp. Ther. 199*, 477-482.

Lu, K.H., Chen, H.T., Grandison, L., Huang, H.H. and Meites, J. (1976). Reduced luteinizing hormone release by synthetic luteinizing hormone-releasing hormone (LHRH) in postpartum lactating rats. *Endocrinology 98*, 1235-1240.

Lu, K.H., Huang, H.H., Chen, H.T., Kurcz, M., Mioduszewski, R. and Meites, J. (1977). Positive feedback by estrogen and progesterone on LH release in old and young rats. *Proc. Soc. exp. Biol. Med. 154*, 82-85.

McGeer, E.G., Fibiger, H.C., McGeer, P.L. and Wickson, V. (1971). Aging and brain enzymes. *Exp. Geront. 6*, 391-396.

Miller, A.E., Shaar, C.J. and Riegle, G.D. (1976). Aging effects on hypothalamic dopamine and norepinephrine content in the male rats. *Exp. Aging Res. 2*, 475-480.

Niswender, G.D., Midgley, A.R., Jr., Monroe, J.E. and Reichert, L.E.,Jr. (1968). Radioimmunoassay of rat luteinizing hormone with anti-ovine LH serum and ovine LH-^{131}I*. *Proc. Soc. exp. Biol. Med. 128*, 807-811.

Peluso, J.J., Steger, R.W. and Hafez, E.S.E. (1977). Regulation of LH secretion in aged female rats. *Biol. Repro. 16*, 212-215.

Pepperell, R.J., Bright, M. and Smith M.A. (1977). Serum prolactin levels in normal women and in women with disorders of menstruation. *Med. J. Aust. 1*, 85-89.

Pepperell, R.J., Evans J.H., Brown, J.B., Bright, M.A., Smith, M.A., Burger, H.G. and Healy, D. (1977). A study of the effects of bromocryptine on serum prolactin, follicle stimulating hormone and luteinizing hormone and on ovarian responsiveness to exogenous gonadotropins in anovulatory women. *Br. J. Ob. Gyn. 84*, 456-463.

Reis, D.J., Ross, R.A. and Joh, T.H. (1977). Changes in the activity and amounts of enzymes synthesizing catecholamines and acetylcholine in brain, adrenal medulla, and sympathetic ganglia of aged rat and mouse. *Brain Res. 136*, 465-474.

Rolland, R., DeJohg, F.H., Schellekens, L.A. and Lequin, R.M. (1975). The role of prolactin in the restoration of ovarian function during the early post-partum period in the human female. *Clin. endocr. 4*, 27-38.

Sherman, B.M. and Korenman, S.G. (1975). Hormonal characteristics of the human menstrual cycle throughout reproductive life. *J. clin. Invest. 55*, 699-706.

Sherman, B.M., West J.H. and Korenman, S.G. (1976). The menopausal transition: Analysis of LH, FSH, estradiol and progesterone concentrations during the menstrual cycles of older women. *J. clin. Endocr. 42*, 629-636.

Simpkins, J.W., Mueller, G.P., Huang, H.H. and Meites J. (1977). Evidence for depressed catecholamine and enhanced serotonin metabolism in aging male rats; Possible relation to gonadotropin secretion. *Endocrinology 100*, 1672-1678.

Thorner, M.O., McNeilly, A.S., Hagan, C. and Besser, G.M. (1974). Long-term treatment of galactorrhea and hypogonadism with bromocriptine. *Br. Med. J. 2*, 419-422.

Treolar, A.E., Boynton, R.E., Behn, B.G. and Brown, B.W. (1967). Variation of the human menstrual cycle through reproductive life. *Int. J. Fert. 12*, 77-126.

Tsai, C.C. and Yen S.S.C. (1971). Acute effects of intravenous infusion of 17β-estradiol on gonadotropin release in pre- and post-menopausal women. *J. clin. Endocr. Metab. 32*, 766-771.

Ungerstedt, U., Ljungberg, T., Hoffer, B. and Siggins, G. (1975). Dopaminergic supersensitivity in the striatum. *Adv. Neurol. 9*, 57-65.

van der Shoot, O. (1976). Changing pro-oestrous surges of luteinizing hormone in aging 5-day cyclic rats. *J. Endocr. 69*, 287-288.

PEPTIDES IN PARKINSON'S DISEASE

ANDRÉ BARBEAU

Department of Neurobiology
Clinical Research Institute of Montreal
Montreal, Quebec, Canada, H2W 1R7

SUMMARY

Peptides have found a progressively more important role in the biochemistry and physiology of the brain in recent years. Many possess neurological and behavioral actions in addition to their function in endocrinology. This paper reviews the formation and distribution, as well as the neurological action, of peptides probably involved in Parkinson's disease: β-LPH and its derivatives (MSH, $ACTH_{4-10}$, β-endorphin). It also recounts the experience of many authors with prolyl-leucyl-glycine-amide (PLG) in that disease. Finally we propose a new theory of the etiology of Parkinson's disease, based on a postulated deficiency in the important trophic function upon catecholaminergic neurons of A.P.U.D. cells.

PEPTIDES IN THE BRAIN

The last 15 years have seen the extraordinary growth of knowledge in the physiology, pharmacology and biochemistry of monoamines, particularly spurred on by the findings of abnormal dopamine metabolism in Parkinson's disease and the encouraging success of levodopa therapy in that illness.

In parallel with this activity, the field of neuroendocrinology has been steadily growing, leading the exciting new findings. One of the products of these studies is undoubtedly the recognition that a number of peptide hormones act directly upon the brain to affect learning and behavior. Prominent among these active hormones are substances isolated from the hypothalamus which, on the other hand, act as releasing or release-inhibiting factors and, on the other hand, possess independent behavior modifying properties.

A number of well known, and important, functions of the organism are dependent upon the action of peptide hormones: sexual behavior, including female receptivity state and mating, are influenced by gonadotropins; eating can be induced by insulin; thirst and the consequent drinking are inhibited by vasopressin, while motivated drinking is induced by angiotensin; parturition can be triggered or accelerated by oxytocin; it is also well known, particularly in frogs, that melanocyte stimulating hormone (MSH) can induce skin color changes; lipolysis can be induced by both MSH and β-lipotropic hormone (β-LPH). Finally the best studied phenomenon is that of neurosecretion in the hypothalamus-pituitary interphase.

Recent developments in protein and peptide biochemistry have permitted the mapping of a number of new pathways using immunohistochemical morphological explorations. Antisera have thus been raised to luteinizing hormone releasing hormone (LHRH), somatostatin, thyrotropin releasing hormone, oxytocin, vasopressin, neurophysin, substance P, enkephalin, angiotensin II, vasoactive intestinal polypeptide, and many others. There are still problems of specificity with this technique, as well as with methods measuring enzymes, but the future of the approach is great *(Hökfelt, Elde, Fuxe, Johansson,, Ljungdahl, Goldstein, Luft, Efendic, Nilsson, Terenuis, Ganten, Jeffcoate, Rehfeld, Said, Perez de la Mora, Possani, Tapia, Teran and Palacios, 1978)*.

PEPTIDE SYNTHESIS IN THE BRAIN

Most of the active peptides studied are not synthesized *de novo*. Rather they originate from the enzymatic breakdown of much larger molecules. This discovery was made simultaneously for insulin, which derives from proinsulin, by Steiner and Oyer *(1967)* and for β-MSH, which originates from lipotropin, by Chrétien and Li *(1967)*. It has led to important advances in the chemistry of many hormones. For example, it is now known that β-LPH, a peptide with 91 amino acids, is the precursor of β-MSH (sequences 37 to 58) which may not even be present as such in humans, and of γ-LPH (sequence 1-58). A recent study *(Hughes, Smith, Kosterlitz, Fothergill, Morgan and Morris, 1975)* indicates that residues 61 to 65 form the structure of methionine-enkephalin, a pentapeptide which may be the natural ligand for opiate receptors. Other fragments with biological activity are now known as α- and β-endorphins. Celis, Taleisnik and Walter *(1971)* have also demonstrated that the tripeptide proline-leucine-glycine amide resulted from the splitting of oxytocin. Such breakdown is carried out by a number of peptidases which are being isolated in the brain. It may well be that the regional concentrations of such peptidases could be the local factor responsible for specificity of peptide mapping.

Some of the early studies on the behavioral effects of peptide hormones were carried out in hypophysectomized animals by De Wied and his colleagues in Utrecht *(De Wied, Whitter and Lande, 1970)*. The animals had decreased extinction of conditioned avoidance responses (CAR) which were corrected by the administration ACTH, MSH or vasopressin. These effects were not dependent on the presence of adrenals. De Wied is of the opinion that ACTH facilitates learning CAR by affecting motivational processes. He and his colleagues have studied many fractions of the ACTH molecules to find that the most active is composed of a 7 amino acid sequence situated between amino acids 4 and 10 (ACTH 4-10): this sequence is Met-Glu-His-Phe-Arg-Try-Gly. Modifications and substitutions of the sequence have been attempted, but most are inactive except when the methionine is oxidized to the sulfoxide, arginine is replaced by D-lysine and tryptophan by phenylalanine. The resulting peptide is 1,000 times more active than MSH/ACTH 4-10, probably because the substitutions increase resistance to degradation by enzymes. There is evidence that MSH/ACTH 4-10 can increase visual memory on the Benton retention test in man and that the peptide prolongs the pattern of mental alertness on the EEG, by decreasing the duration of α pattern. Melanocyte stimulating hormone (MSH) possesses the same 7 amino acid sequence as MSH/ACTH 4-10 and it has essentially the same behavioral action.

A very strange neurological phenomenon consisting of yawning and stretching crises of muscular hypertonus has been observed in dogs and rats after the intracerebral injection of MSH/ACTH 4-10, MSH and β-LPH *(Izumi, Donaldson and Barbeau, 1973)*, all substances possessing the 7 amino acid sequence Met-Glu-His-Phe-Arg-Try-Gly.

β-LPH DERIVATIVES AND PARKINSON'S DISEASE

Cotzias, Van Woert and Schiffer *(1967)* had shown that the injection to MSH to parkinsonian patients rapidly exacerbated the symptoms of the illness. In subsequent studies we were able to confirm this experience, particularly upon tremor. Later Shuster, Thody, Goolamali, Burton, Plummer and Bates *(1973)* demonstrated that plasma MSH values were elevated in Parkinson's disease. However this finding is still in doubt, mainly because the specificity of MSH determination is not as clear as previously thought. The assay may in fact be measuring β-LPH as well as MSH.

Another part of the β-LPH molecule may be playing an important role in Parkinson's disease. As mentioned before the 61-65 sequence of β-lipotropin has been isolated and chemically characterized and recently shown to possess morphine-like properties. Subsequent reports showed that these peptides administered intraventricularly in rats and mice have an analgesic action. The C-fragment (β-LPH 61-91), designated as β-endorphin has been found to have potent and long-lasting analgesic action when administered intraventricularly in cats, rats and mice. Recently, β-endorphin injected into the cisterna magna, in the periaqueductal gray or in the lateral ventricle in rats, has been demonstrated to produce catalepsy or catatonia in addition to analgesia. However, our own experience indicates that the phenomenon of decreased mobility and motor initiation should be called akinesia rather than catatonia

(Izumi, Motomatsu, Chrêtien, Butterworth, Lis, Seidah and Barbeau, 1977). This behavior was fully reversed by naloxone, a specific antagonist for opiate, L-dopa with a peripheral decarboxylase inhibitor or apomorphine. L-dopa or apomorphine did not reverse the analgesia, while naloxone at least partially reversed both analgesia and akinesia. Indeed, apomorphine (20 mg/kg) fully reversed the akinesia induced by β-endorphin and produced its characteristic stereotyped behavioral effect 7 minutes following drug injection. The complete akinesia reversal effect of apomorphine lasted for about 17 minutes, after which time its efficacy decreased gradually. This observed time course in the action of apomorphine seems to be parallel to the time course of accumulation and disappearance of the drug in the rat brain *(Butterworth and Barbeau, 1975).* These findings, together with the recent report that dopamine release from a rat striatal slice is inhibited by β-endorphin, indicate that the peptide acts preferentially upon dopaminergic neurons, probably at presynaptic sites.

Thus, the above evidence indicates that many peptides derived from the prohormone β-LPH could be involved in the production of some of the symptoms of Parkinson's disease, particularly tremor and akinesia.

THE PRO-LEU-GLY-NH_2 (PLG) STORY

In 1971, Nair, Kastin and Schally *(1971)* synthesized a tripeptide, L-prolyl-L-leucyl-glycine amide (Pro-Leu-Gly-NH_2; PLG) which they claimed had MSH-release-inhibitory (MIF) properties. They called this peptide MIF-I, but we would now prefer to use the initials PLG because subsequent studies by many authors have failed to confirm this hormonal activity. However animal experiments soon revealed that this substance was neurologically active, in that it potentiated the actions of L-dopa and oxotremorine in both intact and hypophysectomized animals *(Plotnikoff and Kastin, 1974).*

Based on these premises, Kastin and Barbeau *(1972)* carried out a number of experiments in Montreal. The first set of studies consisted in the slow intravenous infusion of 20 to 40 mg of PLG in 100 ml of saline over a 30-minute period in 8 patients. Rigidity was improved by an average 20%, while tremor was markedly reduced in 4 of the 8 patients. To our surprise this benefit persisted some 2 or 3 days after the infusion. One week later the same 8 patients were given 30 mg per day of oral PLG for 2 days. Motor performance tests, objectively measured, improved by an average 19% and justified longer trials in 3 patients. The latter received 50 mg/day of oral PLG for a minimum of two months. Again the objective motor performance tests were improved by 30% at the end of the observation peroid. A further 5 patients who were taking L-dopa, but presented with oscillations in performance and dyskinesias were given, on a single occasion, 30 mg of PLG orally. Although a slight decrease in the dyskinesias was noted, no clear-cut potentiation of performance could be observed.

These results were soon partially confirmed by Chase, Woods, Lipton and Morris *(1974)* who found some antiparkinsonian activity with small infusions of the drug, but could not confirm the decrease in dyskinesias. The same year Fischer, Schneider, Jacobi and Maxion *(1974)* in Germany made similar positive observations in 10 patients. They observed simultaneous mood brightening and thought that the effect of PLG was mostly upon mood and motivation. These studies were followed by a 4-month double-blind experiment with gradually increasing oral doses of PLG in 20 parkinsonian patients, followed by 6 months of open observation *(Barbeau, Roy and Kastin, 1976)*. Unfortunately, objective measurements of various parameters did not reveal important differences when the initial and final results were compared. Functional impairment and finger dexterity, to all intents and purposes, remained identical. However, there was a significant downward trend in rigidity and tremor scores for the patients receiving PLG. This corresponds to 20% and 44% decreases respectively, figures of the same order of magnitude as seen in the inital study.

A further study, this time using the intravenous approach already shown to be effective, again demonstrated the efficacy of a single bolus injection of 200 mg PLG in 8 parkinsonian patients *(Barbeau and Kastin, 1976)*. This effect persisted for nearly 6 hours.

Animal studies quoted above had shown that PLG potentiates the L-dopa-induced effect on motility. In a further set of experiments *(Barbeau, 1975)*, we were able to show the same type of potentiation in human parkinsonian subjects. Six of our patients who had been treated with L-dopa for an average of a little over 4 years were chosen for evaluation. After this period of time, the average motor performance score was still significantly improved over the pre-dopa period. On the first day of the experiment, at 10:00 AM, the patients were given their usual 500 mg dose of oral L-dopa. At the same time they received an intravenous injection of 10 ml of NaCl. Performance scores were measured before and hourly after the injection. L-Dopa produced an average 21% further improvement. On the second day, 200 mg of intravenous PLG was substituted for the NaCl. This produced an improvement in performance scores of 44% which lasted for at least 4 hours. In 4 of the 6 patients, and for the first time since the very onset of treatment with L-dopa, performance scores within the normal range were obtained. All patients also noted a marked amelioration in the clarity of their thinking. Similar results were recently obtained by Gerstenbrand, Binder, Grünberger, Kozma, Pusch and Reisner *(1976)* and by our group over more chronic periods *(Gonce and Barbeau, 1977)*.

PLG was ineffective, in our studies *(Barbeau and Kastin, 1976)* in modifying brain levels, turnover or distribution of catecholamines. Neither could we demonstrate inhibition of monoamine oxidase or catecholamine-O-methyl transferase activity by PLG. Furthermore, there was no evidence that PLG acted upon the reuptake mechanism or facilitated the release of catecholamines in an amphetamine-like fashion. Therefore, a presynaptic or metabolic mode of action is unlikely. PLG is still active in animals after hypophysectomy, indicating that the effect upon the brain is probably not through peripheral hormones. Finally we have demonstrated *(Barbeau and Kastin, 1976)* that PLG potentiates the action of apomorphine in reversing the akinesia produced by a bilateral hypothalamic lesion with 6-hydroxydopamine in the rat. This would favour a postsynaptic site of action for PLG.

Recently, we have begun studying in animals and man some analogs of PLG which, in rats at least, appear to be as effective as PLG, but less subject to biological degradation. It is hoped that these new products will prove to be useful against the symptoms of Parkinson's disease.

A PEPTIDERGIC THEORY OF THE ETIOLOGY OF PARKINSON'S DISEASE

The actual cause of Parkinson's disease remains an enigma, despite the numerous studies originating from the discovery of the therapeutic activity of L-dopa. There is still no satisfactory answer to the elementary question: what causes the deficiency in dopamine levels in the basal ganglia and elsewhere? Viruses have been implicated in the etiology of the "post-encephalitic parkinsonism" which followed the epidemic of lethargic encephalitis of 1918-1927. However the true nature of the actual virus involved remains unsettles, despite claims in favor of influenza A *(Gamboa, Wolf, Yahr, Harter, Duffy, Borden and Hsu, 1974)*. In this respect, it is of interest to note that Lycke and Roos *(1969)* have indicated that such viruses induce increases in dopamine turnover in mice brain.

In a recent presentation to the Association for Research in Nervous and Mental Disorders, we proposed a new hypothesis on the etiology of Parkinson's disease *(Barbeau, 1976)*. This hypothesis envisions a primordial role for brain peptides and was thus stated: *"The symptoms of Parkinson's disease are a consequence of localized amine imbalances in the brain, but the basic pathogenic mechanism is a accelerated aging phenomenon resulting from the selective atrophy of the heavily pigmented cells in the brainstem, from whence originate dopaminergic and noradrenergic pathways. This aging mechanism can be accelerated by vascular, infectious, and toxic factors, but it is conditioned by a deficiency in specialized neuroendocrine (A.P.U.D.) cells in the hypothalamus. Parkinson's disease is thus a form of A.P.U.D. cell deficiency syndrome"*
The evidence for the first part of this proposition has been reviewed many times *(Barbeau, 1962; Barbeau, 1973; Barbeau, 1976; Barbeau, Campanella, Butterworth and Yamada, 1975)* and will not be repeated here. The only point to be made is the interesting similarity between the mechanisms of aging and that of parkinsonism. Thus, the brainstem pigmented cells are also involved in aging as shown by a significant lowering of striatal dopamine content

in both diseases *(Finch, 1973)*. In such a situation, the slightest further aggression to the integrity of the nigrostriatal pathway will result in the rapid appearance of extrapyramidal signs and symptoms. In fact, various mechanisms (vascular, infectious, and toxic) are known to further worsen the concentration or turnover of brain catecholamines *(Moskowitz and Wurtman, 1975; Barbeau, Rojo-Ortega, Brecht, Donaldson, Minnich and Genest, 1972)*. The common end result is the acceleration of a random degenerative process in the basal ganglia, which eventually leads to obvious extrapyramidal signs. It is likely that these secondary factors alone could not produce the signs and symptoms of Parkinson's disease unless they were extremely severe or accompanied by evidence of an accelerated aging process.

On the other hand, one of the main characteristics of Parkinson's disease is a deficiency in the decarboxylases necessary for synthesis: dopa and 5-HTP decarboxylase, glutamic acid decarboxylase. Cells equipped with such decarboxylases are easily seen with histofluorescence and readily take up the appropriate precursors. Pearse *(1969)* has classified such cells as belonging to the A.P.U.D. system. The term itself derives from the main characteristics of the cells: fluorogenic *A*mine content (catecholamines, 5-HT) and/or amine *P*recursor *U*ptake (dopa or 5-HTP) with presence of amino acid *D*ecarboxylases. The other main characteristic of these cells, mostly originating from the neural crest, is that they are associated with the secretion of polypeptides, some of which have hormonal actions. The A.P.U.D. cells usually manifest their presence through hypersecretion, such as in the multiple endocrine syndrome, the Zollinger-Ellison's syndrome, and other secreting tumors which have received the unelegant name of "apudomas".

However, in our hypothesis we prepose that both aging and Parkinson's disease would be the result of specific *deficiencies* in the function of specialized neuroendocrine (A.P.U.D.) cells.

These cells could be absent or reduced in number congenitally, or they could be damaged progressively through the cumulative effect of successive aggressions upon this system (toxic, infectious, vascular, or metabolic). It is part of our proposal that the *main function of peptidergic pathways is a trophic modulation of aminergic functions*, pre- or post-synaptically. In a manner similar to nerve growth factor in the peripheral nervous system *(Mobley, Server, Ishii, Riopelle and Shooter, 1977)* these polypeptide producing neurons could permit the growth or maintenance of neurotransmitter producing neurons. Any decrease in this trophic action of peptidergic neurons originating from A.P.U.D. cells, through congenital absence or progressive damage, would result in eventual atrophy and death of the latter neurons (dopaminergic and noradrenergic in Parkinson's disease, GABA-ergic in Huntington's chorea). Preliminary evidence from the studies reviewed above would favor the importance of β-LPH derivatives (MSH,ACTH 4-10, β-endorphin) in this role as regards Parkinson's disease, with substance P and renin-angiotensin in Huntington's chorea.

Our hypothesis leads to a number of physiological and pharmacological experiments and to eventual therapeutic approaches, and is now being actively evaluated in our laboratory.

ACKNOWLEDGEMENTS

The studies from the author's laboratory quoted in this paper were supported in part through grants from the Medical Research Council of Canada (MT-4938), the United Parkinson Foundation and the W. Garfield Weston Foundation.

REFERENCES

Barbeau, A. (1962). The pathogenesis of Parkinson's disease: a new hypothesis. *Can. Med. Ass. J. 87*, 802-807.

Barbeau, A. (1973). Aging and the extrapyramidal system. *J. Am. Geriat. Soc. 21*, 145-149.

Barbeau, A. (1975). Potentiation of L-DOPA effect by intravenous L-prolyl-L-leucyl-glycine amide in man. *Lancet 2*, 683.

Barbeau, A. (1976). Parkinson's disease: Etiological considerations. In: *The Basal Ganglia*, pp. 281-292. Ed. M.D. Yahr. Raven Press, *New York*.

Barbeau, A., Canpanella, G., Butterworth, R.F. and Yamada, K. (1975). Uptake and efflux of ^{14}C- dopamine in platelets: Evidence for a generalized defect in Parkinson's disease. *Neurology 25*, 1-9.

Barbeau, A., and Kastin, A.J. (1976). Polypeptide therapy in Parkinson's disease - A new approach. In: *Advances in Parkinsonism*, pp. 483-487. Eds. W. Birkmayer and O. Hornykiewicz. Editiones Roche, *Basle*.

Barbeau, A., Rojo-Ortega, J.M. Brecht, H.M., Donaldson, J., Minnich, J.L. and Genest, J. (1972). Effect of a magnesium-deficient diet on the striatal content of amines in the dog. *Experientia 28*, 289-291.

Barbeau, A., Roy, M. and Kastin, A.J. (1976). Double-blind evaluation of oral L-prolyl-L-leucyl-glycine amide in Parkinson's disease. *Can. Med. Ass. J. 114*, 120-122.

Butterworth, R.F. and Barbeau, A. (1975). Apomorphine: Stereotyped behaviour and regional distribution in rat brain. *Can. J. Biochem. 53*, 308-311.

Celis, M.E., Taleisnik, S. and Walter, R. (1971). Regulation of formation and proposed structure of the factor inhibiting the release of melanocyte-stimulating hormone. *Proc. Natn. Acad. Sci. 68*, 1428-1433.

Chase, T.N., Woods, A.C., Lipton, M.A. and Morris, C.E. (1974). Hypothalamic releasing factors and Parkinson's disease. *Arch. Neuro. 31*, 55-56.

Chrétien, M. and Li, C.H. (1967). Isolation, purification and characterization of gammalipotropic hormone from sheep and pituitary glands. *Can. J. Biochem. 45*, 1163-1174.

Cotzias, G.C., Van Woert, M.H. and Schiffer, L.M. (1967). Aromatic amino acids and modifications of parkinsonism. *New Engl. J. Med. 276*, 374-379.

De Wied, D., Witter, A. and Lande, S. (1970). Anterior pituitary peptides and avoidance acquisition of hypophysectomized rats. In: *Progress in Brain Research*, Vol. 32, pp. 213-230. Eds. D. De Wied and J.A.W.M. Weinjneu. Elsevier Publishing Co., *Amsterdam.*

Finch, C.E. (1973). Catecholamine metabolism in the brains of aging male mice. *Brain Res. 52*, 261-276.

Fischer, P.A., Schneider, E., Jacobi, P. and Maxion, H. (1974). Effect of melanocyte-stimulating hormone-release inhibiting factor (MIF) in Parkinson's syndrome. *Europ. Neurol. 12*, 360-368.

Gamboa, E.T., Wolf, A., Yahr, M.D., Harter, D.H., Duffy, P.E., Barden, H. and Hsu, K.E. (1974). Influenza virus antigen in post-encephalitic parkinsonism brain. *Arch. Neurol. 31*, 228-232.

Gerstenbrand, F., Binder, H., Grünberger, J., Kozma, C., Push, S. and Reisner, T. (1976). Indusion therapy with MIF (melanocyte inhibiting factor) in Parkinson's Disease. In: *Advances in Parkinsonism*, pp. 456-461. Eds. W. Birkmayer and O. Hornykiewicz. Editiones Roche, *Basle.*

Gonce, M. and Barbeau, A. (1977). Expériences thérapeutiques avec le Propyl-Leucyl-Glycine amide dans la maladie de Parkinson. *Revue Neuro.* (Paris), (in press).

Hökfelt, T., Elde, R., Fuxe, Fl, Johansson, O., Ljungdahl, A., Golstein, M., Luft, R., Efendic, S., Nilsson, G., Terenuis, L., Ganten, D., Jeffcoate, S.L., Rehfeld, J., Said, S., Perez, de la Mora, M., Possani, L., Tapia, R., Teran, L. and Palacios, R. (1978). Aminergic and peptidergic pathways in the nervous system with special reference to the hypothalamus. In: *The Hypothalamus*, pp. 69-135. Eds. S. Reichlin, B.J. Baldessarini and J.B. Martin. Raven Press, *New York.*

Hughes, J., Smith, T.W., Kosterlitz, H.W., Fothergill, L.A., Morgan, A.B. and Morris, H.R. (1975). Identification of two Related pentapeptides from the brain with potent opiate agonist activity. *Nature 258*, 577-579.

Izumi, K., Donaldson, J. and Barbeau, A. (1973). Yawning and stretching in rats induced by intraventricularly administered zinc. *Life Sci. 12*, 203-210.

Izumi, K., Motomatsu, T., Chrétien, M., Butterworth, R.F., Lis, M., Seidah, N. and Barbeau, A. (1977). β-endorphin induced akinesia in rats: effect of apomorphine and α-methyl-p-tyrosine and related modifications of dopamine turnover in the basal ganglia. *Life Sci. 20*, 1149-1156.

Kastin, A.J. and Barbeau, A. (1972). Preliminary clinical studies with L-prolyl-L-leucyl-glycine amide in Parkinson's disease. *Can. Med. Ass. J. 107*, 1079-1081.

Lycke, E. and Roos, B.E. (1969). Some virological and biochemical aspects of the pathogenesis of Parkinson's disease. In: *Third Symposium on Parkinson's Disease*, pp. 16-21. Eds. F.J. Gillingham and I.M.C. Donalson. E. and S. Livingstone Ltd., *Edinburgh*.

Mobley, W.C., Server, A.C., Ishii, D.N., Riopelle, R.J. and Shooter, E.M. (1977). Nerve Growth Factor, Parts 1,2 and 3. *New Engl. J. Med. 297*, 1096-1104; 1149-1158; 1211-1218.

Moskowitz, M.A. and Wurtman, R.J. (1975). Catecholamines and neurologic diseases. Parts I and II. *New Engl. J. Med. 293*, 274-280; 332-338.

Nair, R.M.G., Kastin, A.J. and Schally, A.V. (1971). Isolation and structure of hypothalamic MSH release-inhibiting hormone. *Biochem. Biophys. Res. Comm. 43*, 1376-1381.

Pearse, A.G.E. (1969). The cytochemistry and ultrastructure of polypeptide hormone-producing cells of the APUD series and the embryologic, physiologic and pathologic implications of the concept. *J. Histochem. Cytochem. 17*, 303-313.

Plotnikoff, N.P. and Kastin, A.J. (1974). Pharmacological studies with a tripeptide, prolyl-leucyl-glycine amide. *Arch. int. Pharmocodyn. Ther. 211*, 211-224.

Shuster, S., Thody, A.Y., Goolamali, S.K., Burton, J.L., Plummer, N. and Bates, D. (1973). Melanocyte-stimulating hormone and parkinsonism. *Lancet i*, 463-464.

Steiner, D.F. and Oyer, P.E. (1967). The biosynthesis of insulin and a probable precursor of insulin by a human islet cell adenoma. *Proc. Natn. Acad. Sci. 57*, 473-480.

INFLUENCE OF THE THYROID GLAND ON OVARIAN FUNCTION IN THE AGING RAT

RICHARD F. WALKER and RALPH L. COOPER

Center for the Study of Aging and Human Development
Duke University Medical Center
Durham, North Carolina 27710

INTRODUCTION

There is a dramatic decline in regular ovarian cycles with increasing age in rats. The anovulatory conditions which develope are associated with characteristic changes in the vaginal epithelium and have been termed constant estrus (CE) or repetitive pseudopregnancy (RPP). Several studies have demonstrated that this disruption of ovarian function is primarily due to age-related changes that occur within the brain and pituitary. For example, ovulation can be induced in old female rats following systemic injections of luteinizing hormone (LH), human chorionic gonadotropin (HCG) *(Aschheim, 1965)*, or luteinizing hormone-releasing factor *(Aschheim, 1976)*. In addition reinstatement of regular ovarian and/or vaginal cycles has been observed in response to electrical stimulation of the medial preoptic area (MPOA) *(Clemens, Amenomore, Jenkins and Meites, 1969; Everett, Holzinger, Zeilmaker, Redmond and Quinn, 1970)*, or by systemic injection of L-tyrosine, L-dopa, epinephrine, and monoamine oxidase inhibitors *(Clemens et al., 1969; Quadri, Kledzik and Meites, 1973; Linnoila and Cooper, 1976)*, and low doses of progesterone *(Everett, 1940)*. Specific environmental changes will also reinstate ovarian cycles in both CE and RPP rats. Reduced photoperiods terminate CE *(Everett, 1943 and 1970; Aschheim, 1966)*, as does chronic cold exposure for RPP *(Aschheim and Latouche, 1975)*.

In general, these treatments represent an attempt to replace deficient or missing components within the hypothalamo-hypophyseal-ovarian axis. With the exception of the indirect effects brought about through environmental manipulation, none has tested the possibility that the reproductive system may fail with age as the result of specific changes which may not be immediately involved in the regulation of ovarian function. For example, exposure to long photoperiods stimulates thyroid activity *(Mayerson, 1935; Tixier-Vidal, 1959)*, whereas constant light *(Fiske, 1941)* and thyroxine treatment *(Bruni, Marshall, Dibbet and Meites, 1975)* both restore gonadotropin levels during times when they are experimentally depressed. These observations suggest that thyroid hyperactivity is correlated with increased periods of vaginal estrus. Indeed, it has been reported that the gonadotropin response to the stimulus of constant light requires that the CNS be continually exposed to thyroid hormone *(Hagino, 1971)*. Also, exposure to constant light restores ovarian cycles in the anestrous, underfed rat *(Piacsek and Meites, 1967)*. Alternatively, food deprivation depresses thyroid activity in rats *(Reichlin, 1957)* and in humans *(Croxson, Hall, Kletzky, Jaramillo and Nicoloff, 1977)*. If underfeeding results in endocrine hypofunction which is incompatible with vaginal estrus, then constant estrus may be the result of positive functional changes in the thyroid axis. These thoughts led us to investigate the influence of food restriction on the reproductive system in aging female rats. We also tested the influence of thyroid hormone on reproductive function, since it may affect the aging process via its control over food intake *(Everitt and Porter, 1976)*.

MATERIALS AND METHODS

Animals and Conditions. Female Long-Evans hooded rats, raised in our laboratory under 14-hr light, 10-hr dark photoperiod and at 72° ± 2F, were used in this study. The health of this colony is maintained by strict isolation of the progeny of the original Caesarean derived population. The result is a very low incidence of respiratory and renal disease in these rats, even into old age. Test animals for underfeeding and hormone experiments included RPP (16-19 months old), spontaneous CE (12-14 months old), pre-estrapause (PEP) (10-12 months old), and cycling(YC) females (4-6 months old). The term pre-estrapause is used here to indicate an animal which is approaching the age at which regular ovarian function normally ceases. For this purpose, PEP is defined as that age at which 50% of the laboratory population is still cycling. Under the conditions of this laboratory, this event occurs between 10 and 12 months of age.

The rats were housed individually in hanging wire cages or in groups of three in plastic cages with wood shavings for bedding. Whenever housing conditions were changed, the vaginal smear was observed in order to exclude possible generalized effects of this treatment on ovarian function.

Underfeeding. All females to be placed on restricted diets were housed individually and their *ad libitum* food intake measured daily for ten days. Thereafter, each female was restricted to 60% of the *ad libitum* mean.

Hormone Treatment. Thyroid hormone was administered in the rat's diet. Females to be fed thyroid supplemented diets were first placed on powdered Purina Lab chow and supplied wood blocks for gnawing. This switch, from the standard pellet food, had no effect on body weight, and all animals readily adapted to this method of feeding. Specific diets were supplemented with 0.5 or 5.0% crude thyroid extract (Sigma Chemical Co. St. Louis, Mo.) or propylthiouracil (PTU) 0.2%. In addition, PTU suspended in sodium alginate was also injected subcutaneously (1.0 mg/kg) in certain animals. Control animals for PTU injections received daily injections of the vehicle only.

Brain Stimulation. Stainless steel, epoxy-coated electrodes insulated to within 0.5 mm of the tip were stereotaxically implanted into the medial preoptic area. Electrochemical stimulation was accomplished by passing 90 μA anodal DC current for 60 seconds.

Vaginal Smears. In all females, vaginal smears were observed daily commencing at least two weeks prior to experimental treatment and continuing for until at least 30 days following termination of treatment. Ovulation was confirmed in selected females by ovariectomy on the morning of vaginal estrus. The ampulla were inspected for the presence of ova and the number shed recorded. In addition all ovaries were examined histologically.

The number of animals in each experimental group varied and are reported under the specific treatments shown in Table 1.

Specific Experiments. Two groups of experiments were conducted in this study. In the first group, an attempt was made to reinstate vaginal and/or ovarian cycles in aged spontaneous CE or RPP females, as well as in young CE, androgenized females, by underfeeding or experimental manipulation of the animals' thyroid levels. In the second group of experiments similar treatments were used in an attempt to disrupt regular ovarian function in young females. These experimental treatments are outlined in Table 1. The appropriate control groups will be discussed in the Results section.

RESULTS

Underfeeding, Thyroid and PTU Treatment in Anovulatory Females. Underfeeding resulted in a reinitiation of vaginal cycling in every 14-16 month old CE female. The latency to renewal of vaginal cycling after dietary restriction was dependent upon the duration of time that the animal had previously spent in CE. Figure 1 shows this relationship. Representative rats, 388 and 404, promptly began cycling during the weight loss phase, immediately after food restriction. Long-term CE animals, however, required more prolonged periods of restricted food intake to reinitiate cycling. In many cases, cycling began after body weight had stabilized. While the pattern of cycling was generally irregular, the greatest regularity occurred in those animals with a short prior history of CE. They displayed more 4 and 5 day cycles than those that had been in CE longer. The irregular cycles were characterized by extended cornified periods while the diestrous interval was normal. This effect is unlike that seen in YC rats where underfeeding resulted in

TABLE 1. EXPERIMENTAL GROUPS AND TREATMENT

A. UNDERFEEDING, THYROID AND PTU TREATMENT TO ANOVULATORY FEMALES

Group	Vag. sm.	N	Age	Treatment*
1	CE	6	14-16 mo	Underfed
2	RPP	4	16-19 mo	Underfed
3	RPP	4	16-19 mo	*ad lib* + thyroid 0.5%
4	CE	4	14-16 mo	*ad lib* + PTU
5	A-CE**	8	4-6 mo	Underfed

B. UNDERFEEDING THYROID AND PTU TO REGULAR CYCLING FEMALES

Group	Vag. sm.	N	Age	Treatment
1	cycling	8	4-6 mo	Underfed
2	cycling	8	4-6 mo	Underfed + PTU
3	cycling	4	4-6 mo	Underfed + thyroid 5%***
4	cycling	6	4-6 mo	ad lib + thyroid 0.5%
5	PEP	6	9-10 mo	thyroid PE + MPOA stim.

* *Treatment periods varied according to group.*

** *Females were neonatally androgenized on day 5 with a subcutaneous injection of testosterone propionate (1.25 mg/kg, suspended in sesame oil).*

*** *Thyroid was added to the diet of the underfed females after a prolonged leukocytic smear was observed.*

reduced cornification with prolonged diestrous intervals. Despite the time differences in the latency to vaginal cycling, ovariectomy of representative females with both long and short CE histories revealed that ovulation had occurred ($\bar{x}$ ova + 10.1 ± .8). In addition, histological examination of ovaries from these previously CE underfed rats revealed new corpora lutea. After realimentation, periods of vaginal cornification characteristically lengthened, and most animals returned to CE after approximately one month. At this time, body weights were not significantly different than those

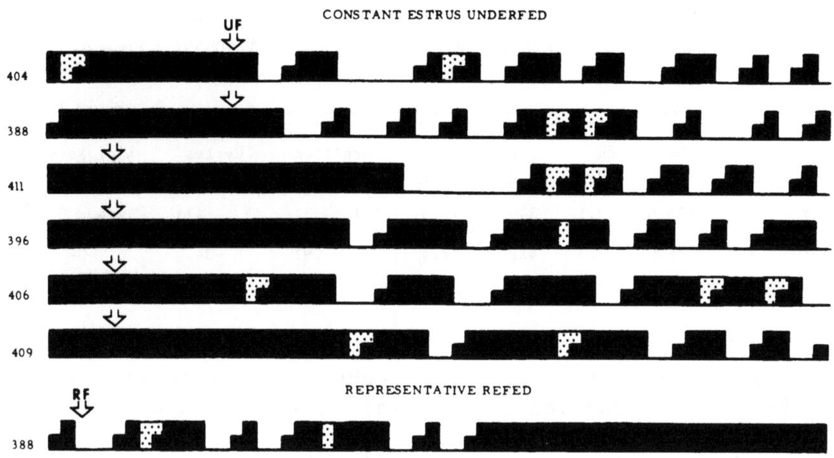

Figure 1. Reinstatement of vaginal cycles in constant estrus rats subsequent to food restriction. UF-day; underfeeding begun; RF-day: Underfeeding discontinued.

__ Diestrus; ■ Proestrus; █ Estrus; ⁞ Leukocytic smear without ovulation.

observed before underfeeding. No rebound or compensatory hyperphagia, nor excessive weight gain, followed refeeding in any animal.

Underfeeding had no effect on the vaginal cytology of the RPP rats. RPP females were responsive to thyroid feeding. *Ad libitum* feeding of thyroid supplemented diet (0.5%) resulted in the prompt induction of an ovulatory cycle in most animals tested (Figure 2). Continued feeding resulted in more vaginal cycles than previously observed for an equivalent period prior to treatment. Both dietary

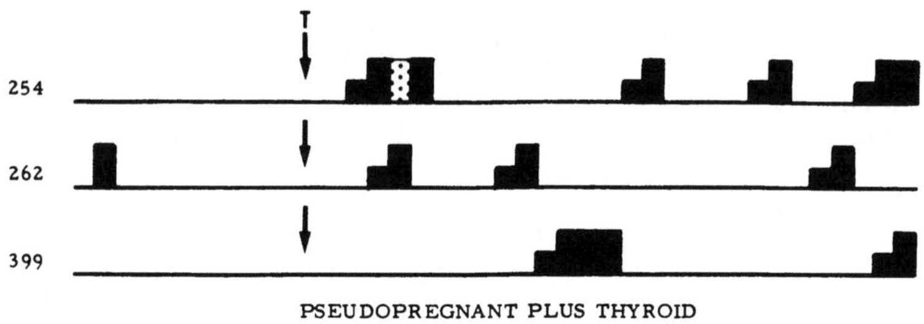

Figure 2. Reinstatement of cycling in representative RPP rats when fed tyroid supplemented diets (N is 4). T: thyroid added to diet. See figure 1 for definition of symbols.

TABLE 2. EFFECT OF THYROID SUPPLEMENTED DIET ON VAGINAL CYCLING IN RPP RATS. CONSUMPTION OF FOOD WAS UNRESTRICTED IN THESE ANIMALS.

Group	N	Beginning Body Weight	Terminating Body Weight	Cornified Days	Cycles	% of Normal Cycles
Thyroid	4	438.6 22.1	340.0 16.5	4.7 .31	3 .471	44.4 6.99
			p .01	p .05	p .001	p .001
Control	4	417.0 18.3	422.3 18.4	2.0 .91	-0-	-0-

thyroid supplementation and underfeeding resulted in a body weight loss. However, the effect of body weight loss *per se* was less influential on the function of the ovarian axis than was dietary thyroid supplementation (Table 2).

When PTU was added to the *ad libitum* diet, irregular vaginal cycles eventually appeared in all CE females. However, this result was seen only after a long duration and was commensurate with a weight loss in these animals. Thus, whether the irregular vaginal cycles that occurred was due to PTU, or weight loss *per se*, or a combination of these two factors, could not be determined.

Perinatally androgenized CE rats did not cycle after food restriction. Instead, leukocytes replaced cornified cells in the vaginal smear after about 10 days of underfeeding. The underfed rats were observed for 21 days, after which it was concluded that caloric restriction is ineffective in stimulating cycles in androgenized CE rats.

Effect of Underfeeding, Thyroid and PTU Treatment on Regular Ovarian Function. Rapid disruption of vaginal cycling did not occur as the result of underfeeding in young female rats; however, definite prolongation of the diestrous interval did occur in the test group as a whole. Treatment with PTU augmented the effect of underfeeding, resulting in a greater depression of cycling in the chemically thyroidectomized animals. Comparison of the effects of chemical thyroidectomy and underfeeding are shown in Figure 3. Figure 4 shows a characteristic thyroid gland from a PTU-treated underfed rat. Hypertrophy of the gland with colloid depletion and follicular collapse occurred in all cases, indicating some continued function of the pituitary thyroid axis, despite severe food restriction. There was significantly greater weight loss in the underfed control group, indicating that thyroid influence over the reproductive axis is greater during underfeeding than absolute body weight (Table 3).

Long-term underfeeding of YC rats resulted in persistent leukocytic vaginal smears; however, a cycle was induced in all such animals, promptly after the inclusion of 5.0% thyroid extract in their daily food ration (Figure 5). This treatment was also accompanied by excessive weight loss; hence, the effect of continued thyroid feeding could not be observed for fear that the rats would die from inanition.

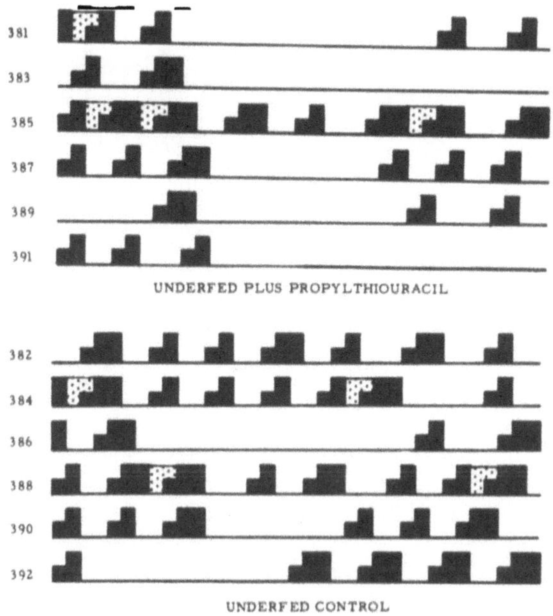

Figure 3. Vaginal cytological alterations following good restriction and/or propylthiouracil injections in normally cycling rats. See figure 1 for definition of symbols.

In both YC and PEP females, the addition of 0.5% thyroid to the *ad libitum* diet resulted in an increase in the number of days a cornified vaginal smear persisted and a decrease in the total number of vaginal cycles observed (Table 4). During the period of treatment, 65% of the PEP rats became CE, while all the control (nonsupplemented) animals continued to cycle (Figure 6). Thyroid treated YC rats returned to normal cycling when thyroid was withdrawn; however, restoration of cycles was often preceeded by prolonged diestrous periods. In contrast, 40% of the PEP females

TABLE 3. INFLUENCE OF UNDERFEEDING AND/OR CHEMICAL THYROIDECTOMY ON CYCLING IN ADULT, SEXUALLY MATURE RATS.

Group	N	Beginning Body Weight	Terminating Body Weight	Cycles During Test	Diestrous Interval	Weight Loss %
Thyroid X	6	292.5 ± 5.06	264.6 ± 4.34	3.8 ± .549 $p<.05$	2.9 ± .261 $p<.05$	9.03 ± .107 $p<.05$
Control	6	300.6 ± 6.15	256.5 ± 4.15	5.7 ± .491	2.3 ± .180	14.68 ± .313

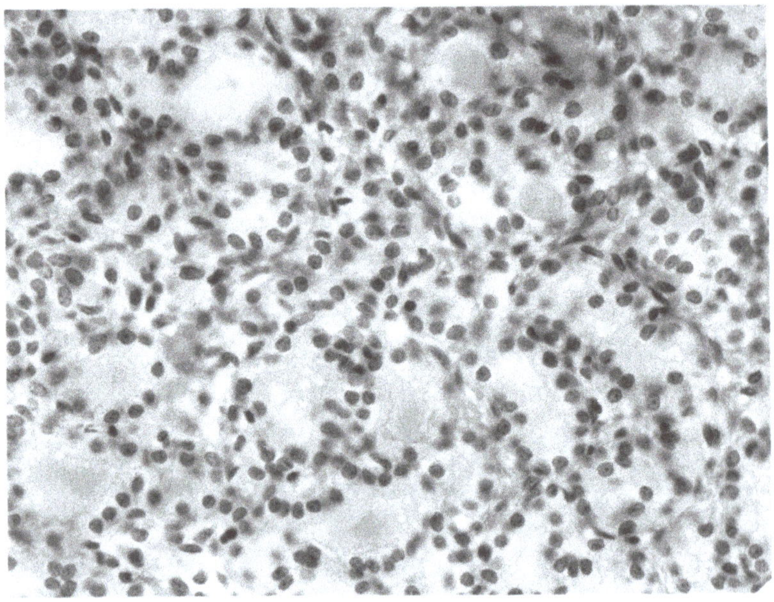

Figure 4. Effect of daily PTU injections in cycling underfed rats. Note the extensive loss of colloid and the resultant change in follicular morphology.

remained CE throughout the 30 day period following treatment withdrawal.

Electrochemical stimulation of the medial preoptic area of YC-thyroid induced persistent estrous rats, restored cycling, and minimized the effect of further tyroid feeding on vaginal cornification (Figures 7 and 8).

TABLE 4. EFFECT OF UNRESTRICTED CONSUMPTION OF THYROID SUPPLEMENTED FOOD IN CYCLING PEP RATS.

Group	N	Beginning Body Weight	Terminating Body Weight	Cycles	Cornified Days	% Change Body Weight
Thyroid	6	339.5 ± 11.1	282.7 ± 8.55 $p<.01$	1.3 ± .273 $p<.05$	14.7 ± .612 $p<.01$	17.3 ± .835 $p<.01$
Control	5	347.1 ± 9.3	341.3 ± 12.11	4.0 ± .942	4.6 ± .546	2.6 ± .177

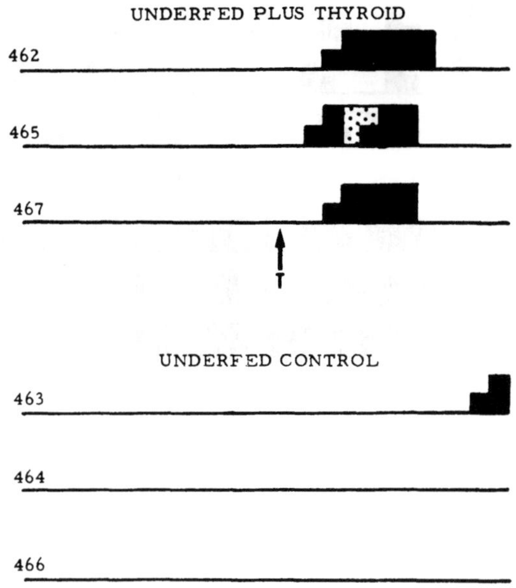

Figure 5. Influence of thyroid extract when given with the restricted food rations of anestrous underfed rats. See figure 1 for definition of symbols. T: day thyroid feeding initiated.

DISCUSSION

A basic premise of many studies which deal with reinitiation of ovarian cycles in old, CE rats is that this condition is due to a deficiency in neurotransmitter control of those neural systems involved in cyclic anterior pituitary function *(Meites, Huang and Simpkins, 1978)*. Indeed age-related deficiencies in hypothalamic catecholamines have been reported *(Finch, 1978)*, and this condition probably accounts for some, if not all, of the senile functional deviations of the aging ovary. The reinstatement of cycling in CE rats by underfeeding is consistent with these suggestions, and also provides a potential mechanism as to how the deficiencies may occur. Pituitary and plasma gonadotropin levels decrease in young animals, and vaginal smears become predominantly leukocytic, when they are underfed *(Piacsek and Meites, 1967)*. That comparable responses, as evidenced by vaginal and ovarian changes, did not occur in CE rats, but instead, quasi-normal function returned, implies that endocrine hyperfunction may be causing the condition, and the "braking" effect of underfeeding restores control. Starvation has been recognized as a method to depress thyroid function for many years *(Reichlin, 1957)*, while exposure to constant light stimulates it *(Tixier-Vidal, 1959)*. The latter treatment also

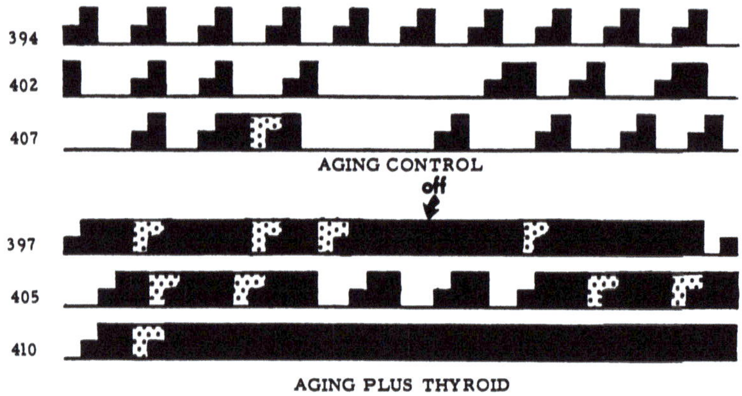

Figure 6. Induction of persistent or constant vaginal cornification by the unrestricted feeding of thyroid supplemented diets to PEP rats. Off indicates withdrawl of thyroid supplementation. See figure 1 for definition of symbols.

elevates pituitary prolactin levels *(Relkin, Adachi and Kahan, 1972)* and induces persistent estrus *(Browman, 1937)*, a condition which requires continuous exposure of the CNS to thyroid hormone *(Hagino, 1971)*. A wealth of information attests to the dependence of the reproductive axis upon the thyroid for normal function. Recent information suggests that the influence of thyroid hormone on

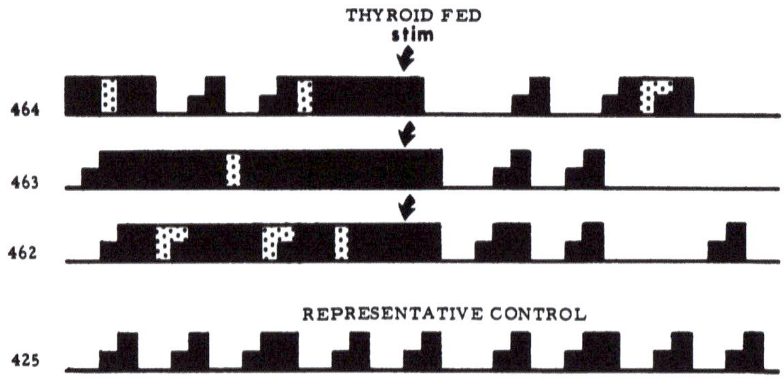

Figure 7. Effect of electrochemical stimulation (90 μA; 60 sec; anodal current) of medial preoptic area on vaginal cytology in thyroid induced persistent estrous rats.

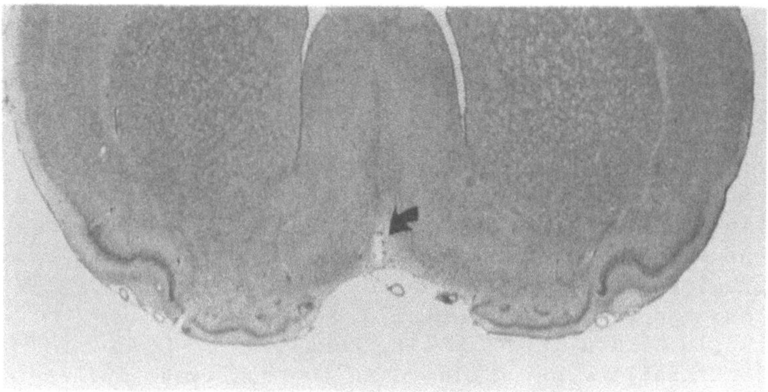

Figure 8. Site of stimulation which reinstated cycles in thyroid induced persistent estrous rats. Lesion resulting from the 90 μA applied current indicated by the arrow.

pituitary gonadotropin content is complex. Aranda, Hervas, Morreale de Escobar and Escobar del Rey *(1976)* showed that low doses of tri- and tetra-iodothyronine, when given to thyroidectomized rats, restored depressed LH levels to normal, while pharmacological doses did not produce any increase in the pituitary hormone. Furthermore, daily increases in the secretion of LH in ovariectomized rats were further enhanced by thyroidectomy *(LaRochelle and Freeman, 1974)*, while thyroxine strongly attenuated the pulsatile discharges of LH in ovariectomized animals *(Freeman, LaRochelle and Moore, 1975)* and reduced LH release by depressing the response of the arcuate nucleus-median eminence area to stimulation *(Freeman, LaRochelle and Moore, 1976)*. Pharmacological doses of thyroxine also decrease the rate of synthesis of brain norepinephrine *(Prange, Meek and Lipton, 1970)*, increase the activity of monoamine oxidase, leading to norepinephrine reduction in the hypothalamus, pons, medulla and midbrain *(Rastogi and Singhal, 1976)*, and reduce the activity of tyrosine hydroxylase, the rate limiting enzyme in catecholamine bio-synthetic pathways, in the median eminence *(Kizer, 1975)*. Coincidentally, such changes could account for the depressing influence of high doses of thyroxine on LH previously reported. Conversely, hypothyroidism increases brain norepinephrine synthesis in the rat *(Lipton, Prange, Dairman and Udenfriend, 1968)*. Underfeeding, therefore, through its effect on the thyroid axis, could result in a partial restoration of brain catecholamines with concomitant reactivation of the senile ovary. Considering these facts, the primary cause of spontaneous CE in rats, may be long-standing exposure to thyroid hormones or a functional hyperthyroidism. This would also explain why thyroxine administration does not induce ovulation in old CE rats *(Gregerman and Bierman, 1974)*, despite the fact that serum thyroxine levels are apparently lower in old than in young rats. While a decline in function of the pituitary-thyroid axis with increasing age has been reported *(Frolkis, Verzhikovskaya and Valueva, 1973)*, others suggest that activity may increase in the older rat *(Gregerman and Crowder, 1963)*. Most probably, the influence of thyroid hormone (s) on its target tissues varies with time, such that comparable amounts produce greater effects in the old than in young animals *(Frolkis et al., 1973)*. Such sensitivity enhancement with age has been reported also for estrogen *(Cooper, 1977)*, norepinephrine and a number of other compounds *(Frolkis, Bezrukov and Sinitsky, 1972)*.

RPP usually occurs later in life than does CE *(Huang and Meites, 1975)*. The inability of underfeeding to reinitiate cycles in the RPP female animals, while thyroid feeding was successful, suggests a terminal condition of general endocrine hypofunction. Previous reports of aging in the thyroid axis, support this contention *(Everitt, 1976)*. Without considering the cause of such age-dependent thyroid change, the absence of thyroid hormone's influence on reproductive cycling would account for the response of RPP rats to the dietary hormones. Speculatively, hypothyroidism may be associated with crowding-induced pseudopregnancy *(Ryan and Schwartz, 1977)*. Such data would further support this hypothesis.

Further evidence to support the idea that reproductive senescence in the rodent may result from altered thyroid relationships comes from the observation that thyroid extract induced persistent estrus in young rats and promoted permanent CE in RPP animals. When care is taken to minimize the catabolic effects of the exogenous hormone on body mass, the response is unequivocal. Cachexia, however, which is characterized by persistent leukocytic smears, obscures the effect. Prior reports of prolonged diestrus associated with chronic thyroid administration *(Weichert, 1930)* are probably due to high dose administration with excessive loss of body weight. The fact that ovarian cycling was restored in thyroid-induced persistent estrous rats by medial preoptic area stimulation, further suggests that the effect of these hormones on ovarian function may be mediated through the central nervous system.

The differential responses of androgenized and aging CE rats to underfeeding indicates that the perinatally testosterone-treated female is a poor model for CE, and probably differs significantly from the aging animal in its neural and endocrine properties.

SUMMARY

The effects of underfeeding and manipulation of the thyroid axis on ovarian function were determined in young and old rats. The depressant effect of reduced food intake on ovarian cycling in young females was potentiated by chemical thyroidectomy, while young anestrous, underfed rats cycled when their diet was supplemented with thyroid extract. These observations indicate that cycling aberrations in underfed rats may occur secondarily to an altered thyroid state.

To determine if thyroid state influences ovarian function in old animals, constant estrous (CE) rats were underfed or chemically thyroidectomized. All underfed rats eventually cycled, while the response to chemical thyroidectomy alone, though still effective, was less dramatic. Realimented CE animals eventually returned to a pattern of constant vaginal cornification. Underfeeding had no effect on ovarian function in old recurrently pseudopregnant females however, these rats responded to thyroid treatment with renewed cycling. 4-6 (YC) 10-12 (PEP) month old females entered a persistent vaginal estrous condition when fed low doses of thyroid extract with their *ad libitum* diet. When the thyroid supplemented diet was discontinued, YC females resumed regular cycling, whereas the vaginal smear in 40% of the PEP rats remained cornified. Cvcling could be restored in YC-thyroid induced CE rats by electrochemical stimulation of the medial preoptic area. These date suggest that senile deviations from normal cycling in the aging reproductive system may be affected by alterations in the thyroid state.

REFERENCES

Aranda, A., Hervás, F., Morreale de Escobar, G. and Escobar del Rey, F. (1976). Effects of small doses of L-thyroxine and triiodo-L-thyronine on pituitary LH content of thyroidectomized rats. *Acta endocr. 83*, 726-736.

Ascheim, P. (1965). La reactivation de l'ovaire des rattes seniles en oestrus permanent au moyen d'hormones gonadotropes ou do la mise a l'obscurite. *C.R. acad. Sci 260*, 2627-2630.

Aschheim, P. (1966). La regulation de la fonction endocrine de l'ovaire chez la ratte Wistar senile. *Proc. 7th Int. Cong. Geront.*, Vienna 2, 105-108.

Aschheim, P. (1976). Aging in the hypothalamic hypophyseal ovarian axis. In: *Hypothalamus, Pituitary and Aging*, pp. 376-418. Eds. J.F. Burgess and A.V. Everitt. Springfield: Charles C. Thomas.

Aschheim, P. and Latouche, J. (1975). Les effets du Sejour au froud sur le cycle ovarien de la ratte et sone age biologique. In: *Problems Actual d' Endocrinologie et de Nutrition*, pp. 95-110. Ed. H.P. Klotz. Paris: Expansion Scientifique Francaise.

Browman, L.G. (1937). Light in its relation to activity and estrous rhythms in the albino rat. *J. exp. Zool. 75*, 375-388.

Bruni, J.F., Marshall, S., Dibbet, J.A. and Meites, J. (1975). Effects of hyper- and hypothyroidism onserum LH and FSH levels in intact and gonalectomized male and female rats. *Endocrinology, 97*, 558-563.

Clemens, J.A., Amenomori, Y., Jenkins, T. and Meites, J. (1969). Effects of hypothalamic stimulation, hormones and drugs on ovarian function in old female rats. *Proc. Soc. exp. Biol Med. 132*, 561-563.

Cooper, R.L. (1977). Sexual receptivity in aged rats: Behavioral evidence for increased sensitivity to estrogen. *Hormones and Behavior 9*, 321-333.

Croxson, M.S., Hall, T.D., Kletzky, O.A., Jaramillo, J.E. and Nicoloff, J.T. (1977). Decreased serum thyrotropin induced by fasting. *J. clin. Endocr. Metab. 45*, 560-568.

Everett, J.W. (1940). The restoration of ovulatory cycles and corpus luteum formation in persistent-estrous rats by progesterone. *Endocrinology 27*, 681-686.

Everett, J.W. (1943). Further studies on the relationship of progesterone to ovulation and luteinization in the persistent-estrous rat. *Endocrinology 32*, 285-292.

Everett, J.W. (1970). Photoregulation of the ovarian cycle in the rat. In: *Collogues internationaux du CNRS, no 172, La photoregulation chez les oiseaus et les mammiferes*, pp. 387-408. Eds. J. Benoit and I. Assenmacher. Parks: CNRS.

Everett, J.W., Holsinger, J.W., Zeilmaker, G.H., Redmond, W.C., and Quinn, D.L. (1970). Strain differences for preoptic stimulation of ovulation in cyclic, spontaneously persistent-estrous and androgen-sterilized rats. *Neuroendocrinology 6*, 98-108.

Everitt, A.V. (1976). The thyroid gland, metabolic rate and aging. In: *Hypothalamus, Pituitary and Aging*, pp. 511-528. Eds. J.A. Burgess and A.V. Everitt. Springfield: Charles C. Thomas.

Everitt, A.V., and Porter, B. (1976). Nutrition and Aging. In: *Hypothalamus, Pituitary and Aging*, pp. 570-613. Eds. J.A. Burgess and A.V. Everitt. Springfield: Charles C. Thomas.

Finch, C.E. (1978). Reproductive senescence in rodents: Factors in the decline of fertility and loss of regular estrous cycles. In: *The aging Reproductive System, Aging*, Volume 4, pp. 193-212. Ed. E.L. Schneider, New York: Raven Press.

Fiske, V.M. (1941). Effect of light on sexual maturation, estrous cycles and anterior pituitary of the rat. *Endocrinology 29*, 187-196.

Freeman, M.E., LaRochelle, F.T., jr. and Moore, R.B. (1975). Thyroid hormone regulation of the pulsatile discharges of luteinizing hormoen in ovariectomized rats. *Endocrinology 97*, 738-743.

Freeman, M.E., LaRochelle, F.T., Jr. and Moore, R.B. (1976). Effect of thyroid status on spontaneous and induced surges of luteinizing hormone. *Endocrinology 99*, 713-719.

Frolkis, V.V., Bezrukov, V.V. and Sinitsky, V.N. (1972). Sensitivity of central nervous structures to humoral factors in aging. *Exp. Geront. 8*, 285-296.

Frolkis, V.V., Verzhikovskaya, N.V., and Valueva, G.V. (1973). The thyroid and age. *Exp. Geront. 8*, 285-296.

Gregerman, R.I. and Bierman, E.L. (1974). Aging and hormones. In: *Textbook of Endocrinology*, pp. 1059-1070. Ed. R.H. Williams, Philadelphia: W.B. Saunders.

Gregerman, R.I. and Crowder, S.E. (1963). Estimation of thyroxine secretion rate in the rat by the radioactive thyroxine turnover technique: influence of age, sex and exposure to cold. *Endocrinology 72,* 383-389.

Hagino, N. (1971). Influence of hypothyroid state on ovulation in rats. *Endocrinology 88,* 1332-1336.

Huang, H.H. and Meites, J. (1975). Reproductive capacity in aging female rats. *Neuroendocrinology 17,* 289-295.

Kizer, J.S. (1975). Endocrine-induced alterations of monoamine metabolism in brain. In: *Anatomical Neuroendocrinology,* pp. 401-408. Eds. W.E. Stumph and L.D. Grant. New York: S. Karger.

LaRochelle, F.T., Jr. and Freeman, M.E. (1974). Superimposition of thyroid hormone regulation on gonadotropin secretion. *Endocrinology 95,* 379-387.

Linnoila, M. and Cooper, R.L. (1976). Reinstatement of vaginal cycles in aged female rats. *J. Pharmac exp. Ther 199,* 477-482.

Lipton, M.A., Prange, A.J., Dairman, W. and Udenfriend, S. (1968). Increased rate of norepinephrine biosynthesis in hypothyroid rats. *Fedn. Proc. 27,* 399 (abstract).

Mayerson, H.S. (1935). Effect of light and of darkness on thyroid gland of the rat. *Am. J. Physiol. 113,* 659-662.

Meites, J., Huang, H.H. and Simpkins, J.W. (1978). Recent studies on neuroendocrine control of reproductive senescence in rats. In: *The Aging Reproductive System,* pp. 1059-1070. Ed. R.H. Williams. Philadelphia: W.B. Saunders.

Piacsek, B.E. and Meites, J. (1967). Reinitiation of gonadotropin release in underfed rats by constant light or epinephrine. *Endocrinology 81,* 535-541.

Prange, A.J., Meek, J.L. and Lipton, M.A. (1970). Catecholamines: diminished rate of synthesis in rat brain and heart after thyroxine pre-treatment. *Life Sci. 9,* 901-907.

Quadri, S.K., Kledzik, G.S. and Meites, J. (1973). Reinitiation of estrous cycles in old constant-estrous rats by central-acting drugs. *Neuroendocrinology 11,* 248-255.

Rastogi, R.B. and Singhal, R.L. (1976). Influence of neonatal and adult hyperthyroidism on behavior and biosynthetic capacity for norepinephrine, dopamine and 5-hydroxytryptamine in rat brain. *J. Pharmac. exp. Ther. 198,* 609-618.

Reichlin, S. (1957). The effect of dehydration, starvation and pitressin injections on thyroid activity in the rat. *Endocrinology 60,* 470-487.

Rilkin, R., Adachi, M. and Kahan, S.A. (1972). Effect of pinealectomy and constant light and darkness on prolactin levels in pituitary and plasma and on pituitary ultrastructure of the rat. *J. Endocr. 54,* 263-268.

Ryan, K.C. and Schwartz, N.B. (1977). Grouped female mice: demonstration of pseudopregnancy. *Biol. Reprod. 17,* 578-583.

Tixier-Vidal, A. (1959). Donées actuelles sur l'influence de la lumiére sur l'activité thyroikienne ches les vertébres. *Annls Endocr.* *20*, 708-723.

Weichert, C.K. (1930). Effect of experimental hyperthyroidism on reproductive processes of female albino rats. *Physiol. Zool* *3*, 461-466.

HYPOTHALAMIC-PITUITARY-OVARIAN INTERACTIONS DURING REPRODUCTIVE SENESCENCE IN THE RAT

M.M. WILKES,[1] K.H. LU,[2] S.L. FULTON, and S.S.C. YEN

Department of Reproductive Medicine, School of Medicine
University of California, San Diego
La Jolla, California 92093

ABSTRACT

The neuroendocrine status of Long-Evans female rats was evaluated at several key stages of reproductive senescence. Young (4-8 mo), middle-aged (10-14 mo) and old (24-30 mo) animals were studied according to reproductive state. The reproductive states studied were (1) regularly cycling, (2) constant estrus and (3) pseudopregnant, as determined by vaginal smear cytology. Neuroendocrine parameters at the levels of the hypothalamus, pituitary and steroid-producing organs were compared between each group. DA[3], E and NE concentrations in the median eminence of the hypothalamus were determined by a highly sensitive radioenzymatic assay. LRF content in the median eminence was measured by radioimmunoassay. Circulating levels of LH, FSH, PRL and six steroids were determined. Changes in hormone and neurotransmitter concentrations were demonstrated in association with the various stages of reproductive

[1] Rockefeller Foundation Research Fellow in Reproductive Endocrinology.

[2] Present Address: Department of Obstetrics and Gynecology, School of Medicine, Center for Health Sciences, University of California, Los Angeles, Los Angeles, California 90024.

[3] Abbreviations used: norepinephrine (NE); epinephrine (E); dopamine (DA); gonadotropin releasing factor (LRF); prolactin (PRL); prolactin inhibiting factor (PIF); estradiol (E_2); estrone (E_1); progesterone (P); 20α-hydroxyprogesterone (20α-OH-P); testosterone (T); and androstenedione (A); human choronic gonadotropin (HCG)

senescence and with age advancement. These changes involved the hypothalamic, pituitary and steroid systems. NE content in the median eminence, FSH in serum and circulating androstenedione were all significantly increased in middle-aged, cyclic rats prior to the onset of senescent anovulation. DA concentration in 24 mo. old constant estrous rats (30.7±7.7 pg/μg, N=6) and in 30 mo. old pseudopregnant rats (27.5±7.1 pg/μg, N=6) was significantly reduced compared to young (6 mo. old), cyclic controls on proestrous (55.0±4.7 pg/μg, N=12). This DA reduction was associated with a 3-fold increase in circulating prolactin. The results are discussed in terms of a regulatory cascade model of female reproductive senescence *(Finch, 1976)*.

INTRODUCTION

A decline in fertility and subsequent anovulation at the time of midlife are phenomena common to rodents, humans and a variety of other mammals *(Talbert, 1968)*. Among the earliest manifestations of reproductive senescence is an increased incidence of irregular reproductive cycles occurring at 10-12 months of age in rats *(Ingram, 1959; Mandl and Shelton, 1959; Aschheim, 1961; 1965)* and in humans in the 6-8 year period immediately prior to the menopause *(Treolar, Boynton, Benn and Brown, 1967)*.

In rats the onset of senescent anovulation, termed constant estrous, is presaged by a pattern of sporadic estrous cycles characterized by delayed ovulation and prolonged vaginal cornification *(Clemons, Amenomori, Jenkins and Meites, 1969)*. The state of constant estrous may endure in the rat for a year or longer and is generally followed by a period of repeated pseudopregnancies lasting several months *(Huang and Meites, 1975)*. In addition, there may be a terminal phase of anestrous in the rat marked by vaginal atrophy *(Huang and Meites, 1975)*.

It has been apparent for some time that diminished ovarian capacity *per se* cannot account for the extinction of reproductive cycles during aging. Thus, 80% of ovaries transplanted from constant estrous *(Peng and Huang, 1972)* or pseudopregnant *(Ascheim, 1965)* rats into young, cycling recipients were capable of initiating and maintaining normal estrous cycles. Indeed, the constant estrous ovary contains many normal oocytes and multiple follicles *(Block, 1952; Novak, 1970; Huang and Meites, 1975)*. The ovary of the old, pseudopregnant rat contains corpora lutea and few follicles *(Huang and Meites, 1975)*.

The importance of neuroendocrine mechanisms in reproductive senescence has been appreciated for some time *(Clemens et al. 1969)*. Temporary resumption of estrous cycles can be achieved by the administration of a variety of CNS-acting compounds including epinephrine, L-DOPA, iproniazid *(Quadri, Kledzik and Meites, 1973; Huang and Meites, 1975; Linnoila and Cooper, 1975)* or the dopamine agonist lergotrile mesylate *(Clemens and Bennett, 1977)* to old constant estrous or psuedopregnant rats. It has been proposed that changes in the type or quantity of hypothalamic catecholamine output participate in the regulation of reproductive senescence *(Quadri et al., 1973; Shaar, Euker, Riegle and Meites, 1975; Finch, 1976; Simpkins, Mueller, Huang and Meites, unpublished)*. Nonetheless, changes in catecholamine output cannot be viewed apart from changes in circulating pituitary and steroid hormone concentrations. For instance, it has been demonstrated that increases in both serum FSH and prolactin occur in old constant estrous and pseudopregnant rats *(Shaar et al., 1975; Clemens and Meites, 1971)*. By comparison, postmenopausal women experience both an increase in serum FSH and a precipitous fall in circulating estradiol level *(Yen, 1977; Tsai and Yen, 1971; Sherman and Korenman, 1975; Judd, Judd, Lucas and Yen, 1976)*. Reduced serum progesterone has been reported in constant estrous rats *(Chan and Leathem, 1977)*.

To account for the biological and hormonal changes attending reproductive senescence, Finch *(1976)* has proposed that a succession of interactive changes in functional output of the hypothalamus, pituitary and ovary underlies reproductive decline. Thus, a chronic alteration in a neuroendocrine set-point, e.g., the concentration or activity of a hypothalamic neurotransmitter, could produce a further chronic change in pituitary hormone secretion. Altered pituitary output may in turn modify ovarian steroidogenesis. Finally, changing steroid levels may initiate further changes in hypothalamic neurotransmitter activity and so propagate a series of alterations in hypothalamo-pituitary-ovarian dynamics which are ultimately incompatible with continued cyclic reproductive function. This model, termed a regulatory cascade *(Finch, 1976)*, issues from the extensive evidence that neural monoamines, pituitary hormones and steroids interact in the maintenance of regular reproductive cycles. For instance, the normal ovulatory gonadotropin surge in rats can be blocked by administration of either antisera against estradiol *(Ferin, Tempone, Zimmering and Van de Wiele, 1969; Ferin, Dyrenfurth, Cowchock, Warren and Van de Wiele, 1974)* or of several antiadrenergic drugs including dibenamine *(Sawyer, Markee and Hollingshead, 1947)* and haloperidol *(Dickerman, Kledzik, Gelato, Chen and Meites, 1974)*. In order to document a regulatory cascade model of reproductive senescence, it is necessary to show that a succession of changes in hypothalamic, pituitary and ovarian output do in fact occur. The present study addresses the issue of what neuroendocrine changes are demonstrable during the various phases of reproductive senescence in the rat. These changes must then be interpreted in terms of reproductive status, age and current knowledge of neuroendocrine regulatory relationships.

MATERIALS AND METHODS

(1) Animals.

Animals were derived from a colony of female Long-Evans rats maintained in this laboratory for over 2 years. Young, cyclic females (2-4 mo) and middle-aged retired breeders (10-12 mo) were acquired from Charles River Breeding Labs. (Wilmington, MA) and maintained at 25°C under a 14:10 hr lighting regime (lights on at 0600 hr and off at 2000 daily). Rats were housed in groups of 4 and provided with a diet of Purina rat chow (Ralston Co.) and tap water *ad libitum*. All animals were in healthy condition.

Patterns of reproductive function were determined by extensive monitoring of vaginal smear cytology in the colony. From the smear patterns four categories of reproductive function could be distinguished: (a) regular cycling, in which animals display successive, uninterrupted 4- or 5-day estrous cycles; (b) irregular cycling, in which periods of prolonged vaginal cornification alternate with sporadic ovulatory cycles; (c) constant estrus, in which vaginal cornification persists a minimum of 15 days or longer; and (d) repeated pseudopregnancy, marked by persistent diestrus smears and the presence of ovarian corpora lutea.

Animals were sacrificed at 1000-1100 hrs. All rats were examined post mortem for ovarian histology and the presence of pathological lesions or endocrinopathies. Hemorrhagic tumors of the pituitary are found frequently in old Long-Evans rats. All animals with pituitary tumors or with palpable mammary tumors were excluded from the present study.

(2) Microdissection of tissue from the median eminence of the hypothalamus.

Rats were killed by decapitation using a guillotine. Brains were removed within 1-2 min of sacrifice, quick-chilled on Dry Ice for 2-4 min, and then frozen at -10°C. Serial coronal sections, 300 microns in thickness, were cut at -10°C in a cryotome. Microdissection of the median eminence followed the "punch-out" methodology of Palkovits *(1973)*. Microdissection needles of 500μm inner diameter were used. Median eminence localization was in accordance with the rat brain atlas of König and Klippel *(1963)*. Tissue pellets were expelled into 75 μl of ice-cold 0.1 N HCl and homogenized immediately by sonication for 3-5 sec. Five μl of homogenate were removed for determination of protein content by the method of Lowry, Rosebrough, Farr and Randall. *(1951)*, and the remainder was frozen until assay for catecholamines and LRF. The mean yield of median eminence protein in young, cyclic controls on proestrus was 28.6 ± 3.5 μg (N=14), and there were no statistically significant changes in median eminence protein content in any of the groups studied.

(3) Radioenzymatic determination of catecholamines.

Tissue homogenates were centrifuged at 30,000 x g for 4 min at 20°C and then placed on ice. Duplicate 10 μl aliquots were removed from the supernatant for measurement of catecholamines. Dopamine (DA), norepinephrine (NE) and epinephrine (E) were determined by a modified single-isotope radioenzymatic assay using catechol-O-methyltransferase (COMT, E.C. No. 2.1.1.6, chromatographically purified from Sigma Chemical Co., Saint Louis, Mo.), similar to that described by Passon and Peuler *(1973)*. Catechol compounds were radiolabelled by enzymatic methylation using ^{3}H-S-adenosyl-methionine (5-15 Ci/mmole from New England Nuclear). The constituents of the methylation reaction were 0.3 M Tris, 20 mM $MgCl_2$, 1 mM EGTA, 0.1 mM dithiothreitol, 200 units/ml COMT, 0.8 μM ^{3}H-S-adenysylmethionine (pH 8.6). Methylation was terminated after 1 hr by addition of 1 M Na_3BO_3(pH 11) to a final concentration of 0.4 M. The reaction products, in a total volume of 100 μl, were extracted at ambient temperature into 1 ml of H_2O-saturated ethyl acetate/redistilled anisole (10:1). The methylated catecholamines, 3-methoxytyramine (3-MT), noremetanephrine (NM) and metanephrine (MN), were back-extracted from the organic phase into 0.2 ml of 0.1 N HCL containing 25 μg each of unlabelled 3-MT, NM and MN as cold carriers *(Cuello, Hiley and Iverson, 1973)*. To facilitate the separation of the three methylated catecholamines by thin layer chromatography, the molecules were chemically acetylated at 20°C. Acetylation was initiated by the addition of 100 μl of acetic anhydride, 250 mg of solid $NaHCO_3$ and 200 μl of distilled H_2O *(Donahue, Osterburg and Finch, in preparation; Waldi, 1962)* and conducted under constant agitation. Acetylated methoxycatecholamines were extracted into 1 ml of H_2O-saturated ethyl acetate and concentrated by evaporation under a stream of N_2 at 40-45°C. Samples were dissolved in 80-100 μl of H_2O-saturated ethyl acetate and applied to silica gel 60-F254 thin layer chromatography plates (EM Laboratories). Plates were developed for 20 min in cyclohexane/chloroform/methanol/glacial acetic acid (15:25:4:1). The plastic-backed plates contained a fluorescent indicator permitting direct visualization of the spots corresponding to the three catecholamine derivatives under UV irradiation. The spots were marked, cut into scintillation vials and immersed in 0.5 ml of 100% ethanol. Five ml of scintillation cocktail (Concifluor from Mallinckrodt) were added, and the samples were counted. This assay system is linear (r=0.99) for all three catecholamines to at least 1 ng. Measurement sensitivity was 2-10 pg for the three compounds. Tissue concentrations were expressed as the mass of catecholamine divided by the protein content of an equivalent volume of homogenate.

(4) Determination of LRF, LH, FSH, PRL and steroids by radioimmunoassay.

Duplicate radioimmunoassay determinations were conducted for LRF on 15 μl aliquots of the supernatant from median eminence homogenates by the method of Nett, Akbar, Niswender, Hedlund and White *(1973)*.

Trunk blood was collected from each rat at the time of sacrifice. LH, FSH and prolactin were measured in the serum by previously described radioimmunoassay methods *(Niswender, Midgley, Monroe and Reichert, 1968; Niswender, Chen, Midgely, Meites and Ellis, 1969; Monroe, Parlow and Midgley, 1968; Daane and Parlow, 1971)*.

Serum concentrations of six steroids were determined: estradiol (E_2), estrone (E_1), progesterone (P), 20α-hydroxyprogesterone (20α-OH-P), testosterone (T) and androstenedione (A). Radioimmunoassay procedures for these steroid measurements have been described

(Tsai and Yen, 1971; Judd and Yen, 1973; Anderson, Hopper, Lasley and Yen, 1976).

(5) Data analysis.

Statistical differences between experimental groups were determined by the Student's group t test and by analysis of variance.

RESULTS

(1) Experimental design.

The experimental design of the present sutdies is predicated upon a comparison of the neuroendocrine status of rats at several key points during reproductive senescence. Comparisons were made on the basis of two criteria: (1) age and (2) reproductive status, as defined by vaginal smear pattern. The young (4-8 mo), regularly cycling female rat on the morning of proestrus was chosen as a reference point. Since the incidence of irregular estrous cycles and constant estrous vaginal smears increases sharply in the age range 10-14 months *(Ingram, 1959; Mandl and Shelton, 1959; Aschheim, 1961; 1965)*, two groups of rats at this age were selected, which were either still displaying regular cyclic reproductive function or which had already begun to show constant constant estrous smears. middle-aged animals were compared to old (24 mo) constant estrous females and to old (30 mo) pseudopregnant rats. These old rats exhibit a well-known and marked hyperprolactinemia *(Clemens and Meites, 1971; Shaar et al., 1975; Clemens and Bennett, 1977)* and were selected in part on the hypothesis that this hyperprolactinemia might reflect reduced dopaminergic activity in the hypothalamus *(Clemens and Meites, 1971; Huang, Marshall and Meites, 1976; Finch, 1976)*. Indices of neuroendocrine status employed in the present study were the tissue concentrations of NE, E, DA and LRF in the median eminence of the hypothalamus and serum levels of LH, FSH, prolactin and six steroids (E_2, E_1, P, 20α-OH-P, T and A).

(2) Catecholamines and LRF in the median eminence.

Tissue concentrations of the three catecholamines and LRF are given in Figures 1 and 2 and in Table 1. All animals were sacrificed at 1000-1100 hr to avoid the possibility of circadian variations in hypothalamic neurotransmitter content. All experiments were conducted between December and May. No significant changes in catecholamine and LRF content were observed between young, proestrous controls sacrificed in either December or March. Thus, it is unlikely that seasonal variations in neurotransmitter concentrations contribute to the present results *(Wirz-Justice, 1975)*.

Figure 1 reveals a statistically significant ($P<.05$) increase in NE in middle-aged, regularly cycling rats. This 66% NE increase is not found in constant estrous animals of the same age or in any other group.

Tissue levels of E were significantly increased ($P<.05$) exclusively in old pseudopregnant rats (Figure 1). No other changes in

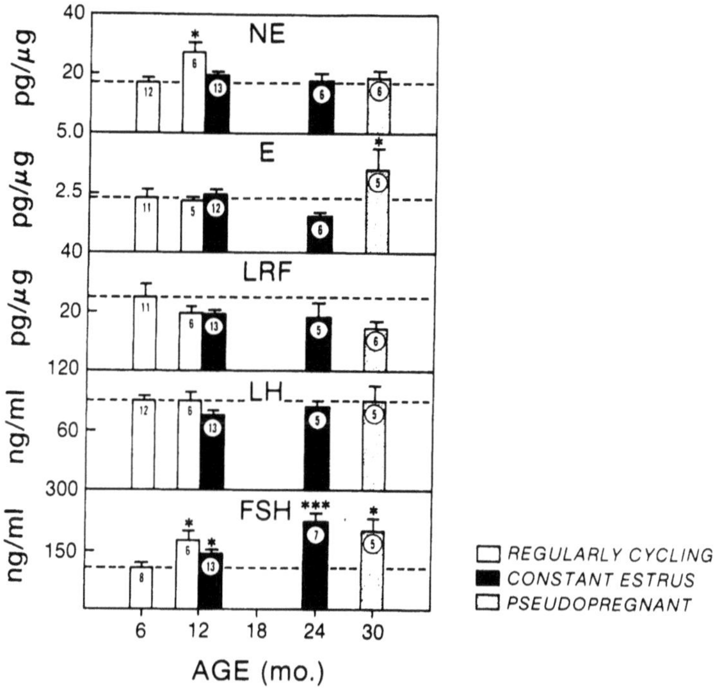

Figure 1. Concentrations of NE, E and LRF in the median eminence and serum gonadotropins.

*All animals were decapitated at 1000-1100 hr. Regularly cycling rats were sacrificed on proestrus. The height of each bar is equivalent to the group mean value. Numbers at the top of the bars are the number of determinations (N). Groups were compared by the Student's t test to the young (6 mo), cycling control group. Statistically significant differences are illustrated as follows: *(P<.05), **(P<.01), ***(P<.001). Criteria for determining reproductive state by vaginal smear cytology are described in Materials and Methods. The age given for each group represents the median value of the age range equal to median ± 1 mo.*

E were demonstrable.

DA concentration was reduced in both old constant estrous and old pseudopregnant rats by two-fold. This reduction was highly statistically significant (P<.01), as shown in Figure 2. Tissue levels of NE and DA in young controls are comparable in magnitude to those reported in other laboratories *(Palkovits, Brownstein,*

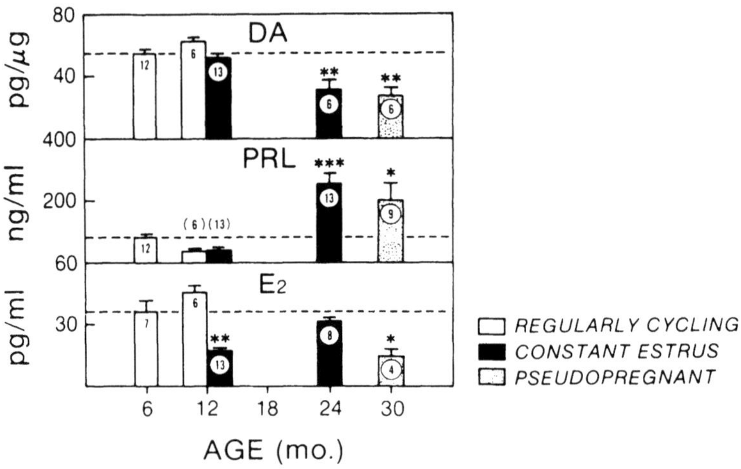

Figure 2. Concentrations of DA in the median eminence and serum PRL and E_2.
Experimental conditions and illustration conventions are the same as in Figure 1.

Saavedra and Axelrod, 1974; Selmanoff, Pramik-Holdaway and Weiner, 1976).

Although there was an apparent tendency toward reduced LRF concentrations during aging (Figure 1), none of the deviations in LRF from control values was statistically significant. LRF levels were comparable in magnitude to those reported by Kobayashi, Lu, Moore and Yen *(1978)*.

(3) LH, FSH and Prolactin.

There were no changes in serum LH concentration in any group tested (Figure 1).

Immunoreactive FSH levels were increased significantly in both middle-aged groups ($P<.05$), in old constant estrous ($P<.001$) rats and in old pseudopregnant animals ($P<.01$). It has been reported previously that FSH is elevated in old, constant estrous rats *(Clemens and Meites, 1971)*. The present data show that this increase occurs relatively early in the lifespan (10-14 mo) and prior to the cessation of regular cycles (Figure 1). However, this FSH elevation

TABLE 1. CONCENTRATIONS OF CATECHOLAMINES AND LRF IN THE MEDIAN EMINENCE AND SERUM PITUITARY HORMONES AND STEROIDS DURING REPRODUCTIVE SENESCENCE.

AGE[a] (mo)	REPRODUCTIVE STATUS[b]	NE (pg/µg)[c]	E (pg/µg)	DA (pg/µg)	LRF (pg/µg)	LH (ng/ml)	FSH (ng/ml)	PRL (ng/ml)	E_2 (pg/ml)	E_1 (pg/ml)	P (ng/ml)	20α-OH-P (ng/ml)	T (pg/ml)	A (pg/ml)
6	Regularly Cycling	16.9 ±2.4 (12)[d]	2.24 ±.48 (11)	55.0 ±4.7 (12)	25.3 ±4.7 (11)	90.5 ±6.0 (12)	104 ±14 (8)	80.9 ±17.8 (12)	36 ±6 (7)	25 ±3 (7)	6.8 ±3.6 (7)	44.3 ±4.4 (7)	70 ±11 (7)	50 ±11 (7)
12	Regularly Cycling	27.0 ±4.4 (6)	2.12 ±.12 (5)	62.1 ±3.9 (6)	19.3 ±3.2 (6)	90.3 ±9.1 (6)	171 ±34 (6)	33.3 ±4.0 (6)	46 ±5 (6)	25 ±2 (6)	8.3 ±2.2 (6)	46.8 ±3.9 (6)	79 ±9 (6)	105 ±14 (6)
12	Constant Estrus	19.5 ±1.2 (13)	2.41 ±.22 (12)	51.6 ±5.0 (13)	19.1 ±2.0 (13)	75.4 ±5.4 (13)	141 ±11 (13)	40.6 ±5.6 (13)	17 ±2 (13)	21 ±2 (13)	3.3 ±.8 (13)	1.2 ±.3 (13)	27 ±4 (13)	31 ±3 (13)
24	Constant Estrus	17.5 ±3.3 (6)	1.50 ±.26 (6)	30.7 ±7.7 (6)	18.0 ±5.6 (5)	85.0 ±5.9 (5)	221 ±21 (7)	249.6 ±45.4 (13)	31 ±4 (8)	34 ±4 (6)	4.6 ±1.0 (8)	1.2 ±.3 (8)	43 ±11 (8)	56 ±11 (8)
30	Pseudo-pregnant	18.5 ±2.8 (6)	3.46 ±.95 (5)	27.5 ±7.1 (6)	14.5 ±2.9 (6)	88.7 ±17.1 (6)	197 ±40 (5)	170.2 ±61.4 (9)	14 ±4 (4)	22 ±5 (4)	27.2 ±8.1 (4)	53.6 ±15.4 (4)	34 ±7 (4)	52 ±10 (4)

[a] Expressed as median age in groups of age range equal to median ± 1 mo.

[b] Criteria used in determining reproductive status by vaginal smear cytology are given in Materials and Methods section.

[c] For catecholamines and LRF, tissue concentration in the median eminence is expressed as mass amount divided by the mass of protein in an equal volume of homogenate.

[d] Numbers in parentheses are equal to the number of determinations (N) made.

first observed in middle-aged, cyclic rats (Figure 1) does not persist throughout the long period of constant estrus. We may compare the FSH concentration of 12 month old rats, which have been in constant estrus for only 2-4 weeks (163 ± 15 ng/ml, N=7) to that in rats showing constant vaginal cornification for 4 weeks or longer (116 ± 6 ng/ml, N=6). FSH in the latter group of relatively long-term constant estrous animals is reduced to the level found in young controls. Hence, while FSH is elevated in middle-aged rats, which have undergone the acute transition to constant estrus, as well as in old, constant estrous animals, this elevation temporarily disappears in middle-aged, long-term constant estrous females. None of the other measured neuroendocrine parameters was significantly different between groups of short-term compared to long-term constant estrous rats.

Figure 2 shows that serum prolactin is significantly elevated in old constant estrous ($P<.001$) and pseudopregnant rats ($P<.05$), in accord with previous observations *(Clemens and Meites, 1971; Clemens and Bennett, 1977)*. No change in circulating prolactin concentration is evident in either middle-aged group (Figure 2, Table 1).

(4) Serum levels of six steroids.

Both estradiol ($P<.01$) and testosterone ($P<.001$) in serum were significantly reduced in middle-aged, constant estrous rats. These two steroids were also significantly lower ($P<.05$) in old pseudopregnant, but not old constant estrous rats (Figure 3, Table 1).

There were no significant changes in estrone (E_1) concentration in any group tested.

Serum progesterone levels in old, pseudopregnant animals were significantly elevated ($P<.05$, Figure 3). Though progesterone was lower in middle-aged constant estrous rats, this change was not significant.

The most striking steroid change observed was a 95% fall in circulating 20α-OH-P in constant estrous rats in either the middle-aged or old groups. The 20α-OH-P reduction was highly statistically significant ($P<.001$).

Significant concentration changes in androstenedione (A) occurred only in the two middle-aged groups. A was increased ($P<.05$) in 12 month old, cycling animals and decreased ($P<.01$) in constant estrous rats of the same age.

DISCUSSION

(1) Neuroendocrine status of middle-aged, cyclic rats.

The advancement in age in cyclic rats between the reproductive prime (2-8 mo) and the age of markedly increased incidence of senescent anovulation (10-14 mo) is attended by significant increases in three neuroendocrine parameters: NE in the median eminence, serum FSH and androstenedione (Figures 1 through 3). The incidence of regular cyclicity approaches zero by the age of 15-17 months in

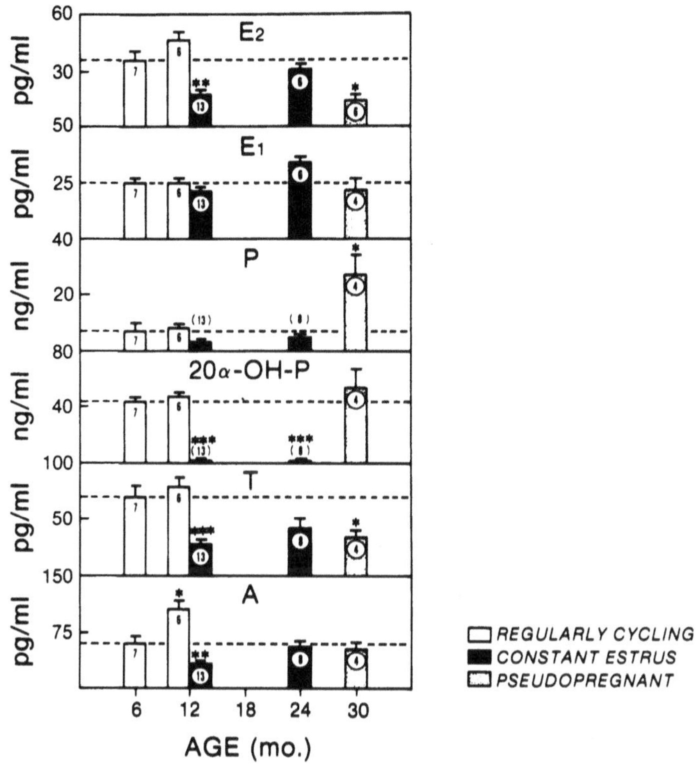

Figure 3. Concentrations of six steroids in serum during reproductive senescence.
Experimental conditions and illustration conventions are the same as in Figure 1.

our colony of Long-Evans females *(Lu, Wilkes and Yen, unpublished observations).* Thus, it may be presumed that this group of middle-aged, cycling rats would have become constant estrous within 2-4 months after these experiments were performed.

The physiological significance of the observed neuroendocrine changes must be assessed with caution. First, at the present time no method is available to determine the functional activity of a neurotransmitter in the median eminence unequivocally. Thus, the present data cannot prove that increased tissue concentration of NE reflects increased neural transmission in NE-containing nerve terminals of the median eminence. Further studies are needed to determine whether NE turnover is also increased in these rats. Unfortunately, the various indices of turnover in current use also do not provide definitive evidence of changes in neural transmission *(Anton-Tay, Anton and Wurtman, 1970; Fernstrom and Wurtman, 1977).* Despite these considerations, it seems likely that the observed NE increase signifies increased activity of this amine. Thus, it has been demonstrated that the concentration of NE increases in both the median eminence *(Chiocchio, Negro-Vilar and Tramezzani, 1976)* and in whole hypothalamus *(Wurtman, Anton-Tay and Anton, 1969)* after castration. The NE increase after castration is associated with increased NE turnover as estimated by uptake of tritiated catechols or increased catecholamine decay rate after administration of α-methyl-p-tyrosine *(Wurtman et al., 1969; Anton-Tay and Wurtman, 1971)* in whole hypothalamus. These changes are concommitant with depletion of LRF in the ME, presumably due to a high release rate *(Kobayashi et al., 1978)*. Thus, at least in the castrate model increased NE concentration in the ME appears to reflect increased functional activity.

Second, increases in immunoreactive FSH do not prove that the functional efficacy of the FSH in circulation is increased. Changes in tissue FSH receptor populations may also contribute to the net functional effect of this hormone during reproductive senescence. Further, the possibility of molecular alterations in the FSH molecules secreted by older animals may change the biological potency or half-life of the hormone. Thus, increased serum FSH could reflect decreased turnover rather than increased FSH secretion. This does not appear to be the case in post-menopausal women. The FSH rise in women is due to increased production and not decreased turnover *(Yen, Tsai, Naftolin, Vandenburh and Ajabor, 1972; Coble, Kohler, Gargille and Ross, 1969)*. Nonetheless, all changes in hormone concentration in the circulating compartment must be viewed in light of possible alterations in receptor populations during aging. Recently, it has been found that the number of LH receptors in the ovary and ovarian capacity to convert androstenedione to estrogens by aromatization are unchanged in old, constant estrous rats *(Erickson, Hsueh, Lu and Yen, unpublished observations)*.

A regulatory cascade model of reproductive senescence postulates that interactive changes in hypothalamic, pituitary and ovarian output may be involved in regulating the onset and course of reproductive decline. Although the relative ovarian contribution to the observed changes in circulating steroids is not known at present, the increases in NE, FSH and A in middle-aged, cyclic rats do encompass the hypothalamic, pituitary and steroid levels of neuroendocrine regulation. Therefore, we may ask whether these three changes may be interactive based on current knowledge of neuroendocrine regulatory relationships. For instance, it is well established that a positive α-noradrenergic input is required to produce the surge of gonadotropins occurring on proestrous day *(Barraclough and Sawyer, 1957; Meyerson and Sawyer, 1967)* or in response to steroid treatment in ovariectomized rats *(Kalra, Kalra, Krulich, Fawcett and McCann, 1972)*.

Multiple lines of evidence suggest that increased noradrenergic activity in the hypothalamus is associated with increased pituitary release of gonadotropins *(Löfstrom, 1977; reviewed by Fernstrom and Wurtman, 1977)*. Thus, increased NE in middle-aged, cyclic rats may contribute to the observed FSH elevation. Alternatively, increased FSH *per se* may produce elevated NE levels by short-loop feedback action. Thus, FSH injections have been found to increase hypothalamic NE turnover *(Anton-Tay et al., 1970)*.

The physiological significance of increased A cannot be readily appraised at present due to a lack of precise information concerning the feedback actions of A on the hypothalamus and pituitary. Also the effects of FSH in the regulation of A production are not well characterized. However, *in vitro* FSH has little effect on ovarian production of A *(Hsueh, Erickson and Yen, unpublished observations)*. A may have potential feedback effects on hypothalamic function by virtue of its aromatization to estrogens in the hypothalamus *(Selmanoff et al., 1976)*. Taken together these considerations are consistent with but do not prove that the observed increases in NE, FSH and A are interactive rather than independent events.

(2) The transition to constant estrous in middle-aged rats.

The transition to the constant estrous state in 12 month old rats is accompanied by a return of NE levels to the control value. As discussed previously, FSH concentration also returns to baseline in 12 month old rats, which have been in constant estrous for at least 4 weeks. The transition to constant estrous results in marked changes in steroid secretion. Statistically significant reductions in concentration of E_2 ($P<.01$), T ($P<.001$), A ($P<.01$) and 20α-OH-P ($P<.001$) are observed during this initial phase of constant estrus. The observed drop in serum E_2 is of particular interest in view of the well-known negative and positive feedback effects of this steroid on the hypothalamo-hypophyseal system *(reviewed by Neill and Smith, 1974)* in the rat, as well as in the human *(Yen, 1977)*. Further, E_2 is required for the production of a normal gonadotropin surge and ovulation *(Ferin et al., 1969)*. The observed two-fold drop in serum E_2 in 12 month old constant estrous rats may in itself be sufficient to explain the failure to ovulate in these animals. Further studies are needed to characterize the steroidogenic capacity of the ovaries of these rats. In 24 month old constant estrous rats, the ovary responds to hCG challenge by increased steroidogenesis either *in vivo (Crumeyrolle-Arias, Scheib and Aschheim, 1976)* or *in vitro (Chan and Leathem, 1977)*. The factors contributing to the reduced steroid secretion in these transitional phase rats remain to be determined

(3) Neuroendocrine status of old (24 mo) constant estrous rats.

The state of constant estrous may persist as long as one year or longer in rats *(Aschheim, 1961; 1965)*. Thus, constant estrous encompasses nearly one third of the rat lifespan. In order to determine the effects of age advancement on the neuroendocrine status of constant estrous rats, 24 month old females were selected. At this age the incidence of constant estrous is still large (around 40%; *Lu, Wilkes and Yen, unpublished observations)* but decreasing, while the frequency of senescent pseudopregnancies is rapidly increasing. Thus, rats of this age may be considered to be in the terminal phase of constant estrous or, alternately, the transition phase to pseudopregnancy.

In these old constant estrous rats, DA concentration in the median eminence is reduced two-fold, and this difference is statistically significant ($P<.01$). Occurring concomitantly with this fall in DA is a 3-fold increase in serum prolactin. The potential physiological significance of reduced DA in the median eminence resides in the well known actions of DA in inhibiting prolactin secretion *(Meites, Nicoll and Talwalker, 1963; Lu, Amenomori, Chen and Meites, 1970)*. In rats, DA can inhibit PRL secretion *in vitro (Macleod, 1969; Birge, Jacobs, Hammer and Daughaday, 1970; Sharr and Clemens, 1974; Takahara, Arimuri and Schally, 1974)* and *in vivo (Blake, 1976)*. The DA antagonist haloperidol increases serum PRL when administered on proestrous morning *(Dickerman et al., 1974)*. Further, the PIF activity of hypothalamic extracts appears to be proportional to catecholamine content *(Schally, Arimura, Takahara, Redding and Dupont, 1974)*. Current understanding of this issue favors the view that DA may be the active physiological PIF. High concentrations of DA have been found both in the median eminence *(Palkovits et al., 1974)* and in pituitary portal blood *(Ben-Jonathan and Porter, 1976; Ben-Jonathan, Oliver, Weiner, Mical and Porter, 1977)*. However, to date it has not been possible to establish an inverse relationship between endogenous dopaminergic activity and the level of prolactin secretion. Viewed in this light, the reduced DA in association with hyperprolactinemia in old constant estrous rats represents strong evidence of an inverse regulatory relationship between these two compounds. However, this conclusion must be further substantiated by turnover measurements. At present it seems likely that the observed fall in DA concentration reflects reduced dopaminergic neural activity. For instance, it has been shown that both DA concentration and turnover are coordinately reduced in the striatum of old, male mice *(Finch, 1973)*. More recently it has been reported that the unit firing activity of dopaminergic neurons of the substantia nigra is directly proportional to DA concentration, as assayed by histochemical fluorescence *(Lichtensteiger, Felix, Leinhert and Hefti, 1976)*.

FSH increases markedly in old, constant estrous rats compared to either young cyclic animals or middle-aged, constant estrous females.

E_2, T and A were substantially reduced in the transition to constant estrous in 12 month old rats, but were increased in old (24 mo), constant estrous rats to approximately the same level as found in young, cyclic controls (Figure 3). It will be of interest to determine what proportion of these steroid increases in old constant estrous rats derive either from peripheral conversion or an adrenal source. In any case, the ovaries of these 24 month old animals are themselves capable of sustaining increased steroidogenesis in response to hCG *in vivo (Crumeyrolle-Arias et al., 1976)* and *in vitro (Chan and Leathem, 1977)*, aromatization of A to estrogens and a full complement of LH receptors.

In contrast to these increases in E_2, T and A, levels of 20α-OH-P remained low relative to controls. It is noteworthy that this sharp reduction in 20α-OH-P was the only observed neuroendocrine change unique to the state of constant estrous. Thus, 20α-OH-P declines at the onset of persistent vaginal cornification (12 mo) and remains low in constant estrous animals as old as 24 month. Yet levels of this steroid return to control values after the onset of senescent pseudopregnancy (Figure 3). The concentration of 20α-OH-P in constant estrous rats was significantly different from that in the other groups both when the data are grouped according to age ($P<0.01$) and to reproductive state ($P<0.01$), as determined by analysis of variance.

It is of interest to consider the possible interactivity of the observed changes in DA, PRL and E_2 in light of the regulatory cascade hypothesis. It seems highly likely that the DA decrease and PRL increase observed in old, constant estrous rats are functionally linked events. Under this interpretation the fall in DA concentration may be considered to be an alteration in neuroendocrine set-point. Hyperprolactinemia may result directly from the DA reduction. We may ask what role E_2 potentially plays in producing the observed changes in DA and prolactin. Although the feedback characteristics of E_2 on the DA-containing nerve terminals of the median eminence are not well characterized, some evidence suggests that E_2 can act to slow DA turnover and reduce DA concentration. Under conditions of high endogenous E_2 stimulation on proestrous day, DA turnover is reduced in the median eminence *(Lofstrom, 1977)*. Recent results indicate that castration in old, constant estrous rats results in a two-fold increase in DA to the same level found in young controls. The administration of silastic implants containing E_2 prevents this DA rise *(Wilkes, Lu and Yen, unpublished observation)*. Marked fluctuations in serum E_2 level occur during the estrous cycle *(Smith, Freeman and Neill, 1975)* in young rats. However, E_2 appears to remain constant at a relatively high physiological level (30 pg/ml) in old, constant estrous rats (Figure 2). If indeed one action of E_2 is to reduce DA activity in the median eminence, then the chronic E_2 elevation in old rats may represent a precipitating factor in the observed changes in DA and PRL.

(4) Neuroendocrine profile of senescent pseudopregnancy.

FSH and prolactin remain high in old, pseudopregnant rats, DA remains low. Circulating P is increased. 20α-OH-P rises dramatically in these rats to the level found in controls (Figure 3).

E in the median eminence is significantly increased ($P<.05$) in these 30 month old pseudopregnant rats. Little is known of the potential regulatory role of this catecholamine in reproductive function. Suggestive of a regulatory role for E are the findings that E is the only catecholamine capable of overcoming pentobarbital blockade of ovulation *(Rubinstein and Sawyer, 1970)* and that E is increased on proestrus *(Wilkes, Lu and Yen, unpublished observation)*.

Nonetheless, much further evidence is needed to substantiate the role of E as a putative regulator in this system.

Both E_2 and T are significantly reduced ($P<.05$) in old, pseudopregnant females (Figure 3). It is of interest that fluctuations in concentration of these two steroids during the various stages of reproductive senescence are not monotonic. Thus, both E_2 and T fall in middle-aged, constant estrous rats, then rise in the 24 month old constant estrous female and finally fall again in the old pseudopregnant animal. This pattern of change clearly suggests that the reduction in serum steroids at the onset of constant estrus does not represent a permanent functional impairment in steroidogenesis.

(5) Regulatory cascade model of reproductive senescence.

The present data indicate that reproductive senescence entails a series of successive alterations in neuroendocrine output. These changes are related to both age and reproductive status. Current understanding of neuroendocrine regulatory relationships does not permit meaningful interpretation of all these changes. Some of the observed changes can be comprehended in terms of established regulatory mechanisms. Thus, the observed inverse relationship between DA and prolactin during reproductive senescence is consistent with the putative role of DA as prolactin inhibiting factor. Physiological and pharmacological evidence suggests that E_2 may play an important role in modulating the DA-PRL system by feedback action.

Further studies will be required to establish whether any of these changes, either singly or in combination, can account for the observed biological decline in reproductive capacity. Also, further studies are needed to confirm whether the observed changes in concentration of neuroendocrine regulatory molecules correspond to changes in functional activity. Lastly, it would be of great interest to learn which changes compromise the adaptive capacity of the animal, as exemplified by the marked increases in incidence of tumors, pathological lesions and mortality during the period after the onset of senescent anovulation.

Taken together, the present data are consistent with, but cannot prove, a regulatory cascade model of reproductive senescence. The alterations in neuroendocrine status during senescence involve the hypothalamic, pituitary and steroid systems. At least some of the observed changes may represent interactions between the hypothalamus, pituitary and ovary.

ACKNOWLEDGEMENTS

The present study was supported by Rockefeller Foundation Grant RF-75029 and by a Rockefeller Foundation Post-Doctoral Fellowship to M.M.W. We thank our colleagues Dr. Mukundan Aiyer, Dr. Aaron Hsueh and Dr. Gregory Erickson for many helpful discussions. We are indebted to Ms. Patricia Gallagher, Mr. Bill Hopper and Mr. Scott Dolin for their excellent technical assistance.

REFERENCES

Anderson, D.C., Hopper, B.R., Lasley, B.L. and Yen, S.S.C. (1976). A simple method for the assay of eight steroids in small volumes of plasma. *Steroids 28*, 179-196.

Anton-Tay, F., Anton, S.M. and Wurtman, R.J. (1970). Mechanism of changes in brain norepinephrine metabolism after ovariectomy. *Neuroendocrinology 6*, 265-273.

Anton-Tay, F. and Wurtman, R.J. (1971). Brain monoamines and endocrine function. In: *Frontiers in Neuroendocrinology* pp. 45, eds. L. Martini and W.F. Ganong. Oxford U. Press, New York.

Aschheim, P. (1961). La pseudogestation â répétition chez les rattes séniles. *C. r. Séanc. Soc. Biol. 253*, 1988-1990.

Aschheim, M.P. (1965). La reâctivation de l'ovaire des rattes séniles en oestrus permanent au moyen d'hormones gonadotropes ou de la mise a l'obscurité. *C. r. Séanc. Soc. Biol. 260*, 5627-5630.

Barraclough, C.A. and Sawyer, C.H. (1957). Blockade of the release of pituitary ovulating hormone in the rat by chloropromazine and reserpine: possible mechanisms of action. *Endocrinology 61*, 341-351.

Ben-Jonathan, N., Oliver, C., Weiner, H.J., Mical, R.S. and Porter, J.C. (1977). Dopamine in hypophysial portal plasma of the rat during the estrous cycle and throughout pregnancy. *Endocrinology 100*, 452-458.

Ben-Jonathan, N. and Porter, J.C. (1976). A sensitive radioenzymatic assay for dopamine, norepinephrine and epinephrine in plasma and tissue. *Endocrinology 98*, 1497-1507.

Birge, C.A., Jacobs, L.S. Hammer, C.T. and Daughaday, W.H. (1970). Catecholamine inhibition of prolactin secretion by isolated rat adenohypophyses. *Endocrinology 86*, 120-130.

Blake, C.A. (1976). Effects of intravenous infusion of catecholamines on rat plasma luteinizing hormone and prolactin concentrations. *Endocrinology 98*, 99-104.

Block, E. (1952). Quantitative morphological investigations of the follicular system in women. *Acta anat. 14*, 108-123.

Chan, S.W.C. and Leathem, J.H. (1977). Aging and ovarian steroidogenesis in the rat. *J. Geront. 32*, 395-401.

Chiocchio, S.R., Negro-Vilar, A. and Tramezzani, J.H. (1976). Acute changes in norepinephrine content in the median eminence induced by orchidectomy or testosterone replacement. *Endocrinology 99*, 629-635.

Clemens, J.A., Amenomori, Y., Jenkins, T. and Meites, J. (1969). Effects of hypothalamic stimulation, hormones, and drugs on ovarian function in old female rats. *Proc. Soc. exp. Biol. Med. 132*, 561-563.

Clemens, J.A. and Bennett, D.R. (1977). Do aging changes in the preoptic area contribute to loss of cyclic endocrine function? *J. Geront. 32*, 19-24.

Clemens, J.A. and Meites, J. (1971). Neuroendocrine status of old constant-estrous rats. *Neuroendocrinology 7*, 249-256.

Coble, Y.D., Kohler, P.O., Gargille, C.M. and Ross, G.T. (1969). Production rates and metabolic clearance rates of human follicle-stimulating hormone in premenopausal and post-menopausal women. *J. clin. Invest. 48*, 359-363.

Crumeyrolle-Arias, M., Scheib, D. and Aschheim, O. (1976). Light and electron microscopy of the ovarian interstitial tissue in the senile rat: normal aspect and response to HCG of "Deficiency Cells" and "Epithelial Cords". *Gerontology 22*, 185-204.

Cuello, A.C., Hiley, R. and Iverson, L.L. (1973). Use of catechol-O-methyltransferase for the enzyme radiochemical assay of dopamine. *J. Neurochem. 21*, 1337-1340.

Daane, T.A. and Parlow, A.F. (1971). Periovulatory patterns of rat serum follicle stimulating hormone and luteinizing hormone during the normal estrous cycle: Effects of pentobarbital. *Endocrinology 88*, 653-663.

Dickerman, S., Kledzik, G., Gelato, M., Chen, H.J. and Meites, J. (1974). Effects of haloperidol on serum and pituitary prolactin, LH, FSH, and hypothalamic PIF and LRF. *Neuroendocrinology 15*, 10-20.

Ferin, M., Dyrenfurth, I., Cowchock, S., Warren, M. and Van de Weile, R.L. (1974a). Active immunization to 17β-Estradiol and its effects upon the reproductive cycle of the Rhesus monkey. *Endocrinology 94*, 765-776.

Ferin, M., Tempone, A., Zimmering, P.E. and Van de Wiele, R.L. (1969). Effect of antibodies to 17β-Estradiol and progesterone on the estrous cycle of the rat. *Endocrinology 85*, 1070-

Ferin, M., Warren, M., Dyrenfurth, I., VAn de Wiele, R.L. and White, W.F. (1974a). Response of rhesus monkeys to LRH [LH-releasing hormone] throughout the ovarian cycle. *J. clin. Endocr. Metab. 38*, 321-337.

Fernstrom, J.D. and Wurtman, R.J. (1977). Brain monoamines and reproductive function. In: *International Review of Physiology, Reproductive Physiology II* pp. 23-55 , ed. R.O. Greep, vol 13. University Park Press, Baltimore.

Finch, C.E. (1973). Catecholamine metabolism in the brains of aging male mice. *Brain Res. 52*, 261-276.

Finch, C.E. (1976). The regulation of physiological changes during mammalian aging. *Q Rev. Biol 51*, 49-83,

Huang, H.H., Marshall, S. and Meites, J. (1976). Induction of estrous cycles in old non-cyclic rats by progesterone, ACTH, ether stress on L-DOPA. *Neuroendocrinology 20*, 21-34.

Huang, H.H. and Meites, J. (1975). Reproductive capacity of aging female rats. *Neuroendocrinology 17*, 289-295.

Ingram, D.K. (1959). The vaginal smear of senile laboratory rats. *J. Endocr. 19*, 182-188.

Judd, H.L., Judd, G.E., Lucas, W.E. and Yen, S.S.C. (1974). Endocrine function of the post-menopausal ovary: Concentration of androgens and estrogens in ovarian and peripheral vein blood.

J. clin. Endocr. Metab. *39*, 1020-1024.
Judd, H.L., Lucas, W.E. and Yen, S.S.C. (1976). Serum 17β-estradiol and estrone levels in postmenopausal women with and without endometrial cancer. *J. clin. Endocr. Metab.* *43*, 272-278.
Judd, H.L. and Yen, S.S.C. (1973). Serum androstenedione and testosterone levels during the menstrual cycle. *J. clin. Endocr. Metab.* *36*, 475-481.
Kalra, P.S., Kalra, S.P., Krulich, L., Fawcett, C.P. and McCann, S.M. (1972). Involvement of norepinephrine in transmission of the stimulatory influence of progesterone on gonadotropin release. *Endocrinology* *90*, 1168-1176.
Kobayashi, R.M., Lu, K.H., Moore, R.Y. and Yen, S.S.C. (1978). Regional distribution of hypothalamic luteinizing hormone-releasing hormone in proestrous rats: Effects of ovariectomy and estrogen replacement. *Endocrinology* *102*, 98-105.
König, J.F.R. and Klippel, R.A.(1963). *The Rat Brain,* Krieger Publishing Co., New York.
Lichtensteiger, W., Felix, D., Lienhart, R. and Hefti, F. (1976). A quantitative correlation between single unit activity and fluorescence intensity of dopamine neurones in zona compacta of substantia nigra, as demonstrated under the influence of nicotine and physostigmine. *Brain Res.* *117*, 85-103.
Linnoila, M. and Cooper, R.L. (1975). Reinstatement of vaginal cycles in aged non-cycling female rats. *Gerontologist* *15*, 30.
Lofstrom, A. (1977). Catecholamine turnover alterations in discrete areas of the median eminence of the 4- and 5-day cyclic rat. *Brain Res.* *120*, 113-131.
Lowry, O.H., Rosebrough, N.J., Farr, A.L. and Randall, R.J. (1951). Protein measurement with the folin phenol reagent. *J. biol. Chem.* *193*, 265-275.
Lu, K.H., Amenomori, Y., Chen, C.L. and Meites, J. (1970). Effects of central acting drugs on serum and pituitary prolactin levels in rats. *Endocrinology* *87*, 667-675.
MacLeod, R.M. (1969). Influence of norepinephrine and catecholamine depleting agents on the synthesis and release of prolactin and growth hormone. *Endocrinology* *85*, 916-923.
Mandl, A.M. and Shelton, M. (1959). A quantitative study of oocytes in young and old nulliparous laboratory rats. *J. Endocr.* *18*, 444-450.
Meites, J., Nicoll, C.S. and Talwalker, P.K. (1963). The central nervous system and the secretion and release of prolactin. In: *Advances in Neuroendocrinology*, pp. 238-277. Ed. A.V. Nalbanov. University of Illinois Press, Urbana, Ill.
Meyerson, B.J. and Sawyer, C.H. (1967). Monoamines and ovulation in the rat. *Acta pharmac. tox suppl* *4*, 18-19.
Monroe, S.E., Parlow, A.F. and Midgley, A.R. (1968). Radioimmunoassay for rat luteinizing hormone. *Endocrinology* *83*, 1004-

Neill, J.D. and Smith, M.S. (1974). Pituitary-ovarian interrelationships in the rat. In: *Current Topics in Experimental Endocrinology*, pp.73-106.Eds. James and Martini. Academic Press, New York.

Nett, T.M., Akbar, A.M., Niswender, G.D., Hedlund, M.T. and White, W.R. (1973). A radioimmunoassay for gonadotropin-releasing hormone (Gn-RH) in serum. *J. clin. Endocr. Metab.36*, 880-885.

Niswender, G.D., Chen, C.L., Midgley, A.R., Meites, J. and Ellis, S. (1969). Radioimmunoassay for rat prolactin. *Proc. Soc. exp. Biol. 130*, 793-797.

Niswender, G.D., Midgley, A.R., Monroe, S.E. and Reichert, L.E. (1968). Radioimmunoassay for rat luteinizing hormone with antiovine LH serum and ovine LH-^{131}I. *Proc. Soc. exp. Biol. 128*, 807-811.

Novak, E.R. (1970). Ovulation after fifty. *Ob. Gyn.36*, 903-910.

Palkovits, M. (1973). Isolated removal of hypothalamic or other brain nuclei of the rat. *Brain Res. 59*, 449-450.

Palkovits, M., Brownstein, M., Saavedra, J. and Axelrod, J. (1974). Norepinephrine and dopamine content of hypothalamic nuclei of the rat. *Brain Res. 77*, 137-149.

Passon, P.G. and Peuler, J.D. (1973). A simplified radiometric assay for plasma norepinephrine and epinephrine. *Analyt. Biochem. 51*, 618-631.

Peng, M.T. and Huang, H.H. (1972). Aging of hypothalamic-pituitary-ovarian function in the rat. *Fert. Steril. 23*, 535-542.

Quadri, S.K., Kledzik, G.S. and Meites, J. (1973). Reinitiation of estrous cycles in old constant-estrous rats by central acting drugs. *Neuroendocrinology 11*, 248-255.

Rubinstein, L. and Sawyer, C.H. (1970). Role of catecholamines in stimulating the release of pituitary ovulating hormone(s) in rats. *Endocrinology 86*, 988-995.

Sawyer, C.H., Markee, J.E. and Hollinshead, W.H. (1947). Inhibition of ovulation in the rabbit by the adrenergic-blocking agent dibenamine. *Endocrinology 41*, 395-402.

Selmanoff, M.K., Pramik-Holdaway, M.J. and Weiner, R.I. (1976). Concentrations of dopamine and norepinephrine in discrete hypothalamic nuclei during the rat estrous cycle. *Endocrinology 99*, 326-330.

Schally, A.V., Arimura, A., Takahara, J., Redding, T.W. and Dupont, A. (1974). Inhibition of prolactin release *in vitro* and *in vivo* by catecholamines. *Fedn. Proc. 33*, 237.

Shaar, C.J. and Clemens, J.A. (1974). The role of catecholamines in the release of anterior pituitary prolactin *in vitro*. *Endocrinology 95*, 1202-1212.

Shaar, C.J., Euker, J.S., Riegle, G.D. and Meites, J. (1975). Effects of castration and gonadal steroids on serum luteinizing hormone and prolactin in old and young rats. *J. Endocr. 66*, 45-51.

Sherman, B.M. and Korenman, S.G. (1975). Hormonal characteristics of the human menstrual cycle throughout reproductive life. *J. clin. Invest. 55*, 699-706.

Smith, M.S., Freeman, M.E. and Neill, J.D. (1975). The control of progesterone secretion during the estrous cycle and early pseudopregnancy in the rat: Prolactin, gonadotropin and steroid levels associated with rescue of the corpus luteum of pseudopregnancy. *Endocrinology 96*, 219-226.

Takahara, J., Arimura, A. and Schally, A.V. (1974). Suppression of prolactin release by a purified porcine PIF preparation and catecholamines infused into a rat hypophysial portal vessel. *Endocrinology 95*, 462-465.

Talbert, G.B. (1968). Effect of maternal age on reproductive capacity. *Am. J. Obstet. Gynec. 102*, 451-477.

Treolar, A.E., Boynton, R.E., Behn, B.G. and Brown, B.W. (1967). Variation of the human menstrual cycle through reproductive life. *Int. J. Fert. 12*, 77-126.

Tsai, C.C. and Yen, S.S.C. (1971a). Acute effects of intravenous infusion of 17β-estradiol on gonadotropin release in pre- and post-menopausal women. *J. clin. Endocr. Metab. 32*, 766-771.

Tsai, C.C. and Yen, S.S.C. (1971b). The effect of ethinyl estradiol administration during early follicular phase of the cycle on the gonadotropin levels and ovarian function. *J. clin. Endocr. Metab. 33*, 917-923.

Waldi, D. (1962). Quantitative Bestimmung von Adrenalinlösungen mit Hilfe der dünnschichtchromatographischen Vergleichsmethode. *Mitt. dtsch. Pharmaz. Ges. 32*, 125-128.

Wirz-Justice, A. (1975). Possible circadian and seasonal rhythmicity in an *in vitro* model: monoamine uptake in rat brain slices. *Experientia 30*, 1240-1241.

Wurtman, R.J., Anton-Tay, F. and Anton, S. (1969). On the use of synthesis inhibitors to estimate brain norepinephrine synthesis in gonadectomized rats. *Life Sci. 8*, 1015-1022.

Yen, S.S.C. (1977). Regulation of the hypothalamic-pituitary-ovarian axis in women. *J. Reprod. Fert. 51*, 181-191.

Yen, S.S.C., Tsai, C.C., Naftolin, F., Vandenberg, G. and Ajabor, L. (1972). Pulsatile patterns of gonadotropin release in subjects with and without ovarian function. *J. clin. Endodr. Metab. 34*, 671-675.

AGE-RELATED CHANGES IN PENILE ERECTIONS AND CIRCULATING TESTOSTERONE IN MIDDLE-AGED MALE RATS

GARY D. GRAY

Department of Physiology
Stanford University School of Medicine
Stanford, California 94143

This symposium has documented a variety of neural and endocrine changes associated with the process of aging. The present chapter focuses on a specific behavioral consequence of these neuroendocrine changes. The behavior is penile erection. It has long been recognized that the erection response in human males declines with increasing age although this was first described systematically by Kinsey, Pomeroy and Martin *(1948)*. Yet, the factors which account for the decline are poorly understood. One possible factor is testicular dysfunction, and innumerable sages, alchemists, and doctors have attempted to restore potency in aging men by administering testicular products, with varying degrees of success.

Our approach to the question of aging effects on penile erection has been to develop an experimental animal model for investigating the erection response, to quantify age-related changes in the behavior, and then to examine the relationship between behavioral changes and alterations in endocrine function. A technique for measuring the erection response in normal male rats has recently been developed in this and other laboratories *(Sachs and Garinello, in press; Davidson, Stefanick, Sachs and Smith, in press)*, based on earlier investigations of erections in spinally-transected or anesthetized animals *(Hart, 1968; Rodgers and Alheid, 1972)*. A male rat is placed on its back, the frontal portion of its body is inserted into a plastic tube, and the back legs are held firmly to keep the animal stationary. The penile sheath is then retracted and held with a wooden rod (Figure 1-top panel). If this position is retained for a few minutes, the animal will typically begin to show erections (Figure 1-bottom panel). These are full erections lasting approximately 1-5 seconds and are comparable in size and duration to erections which occur during normal mating. Some of the erections may even include a pronounced flaring of the glans

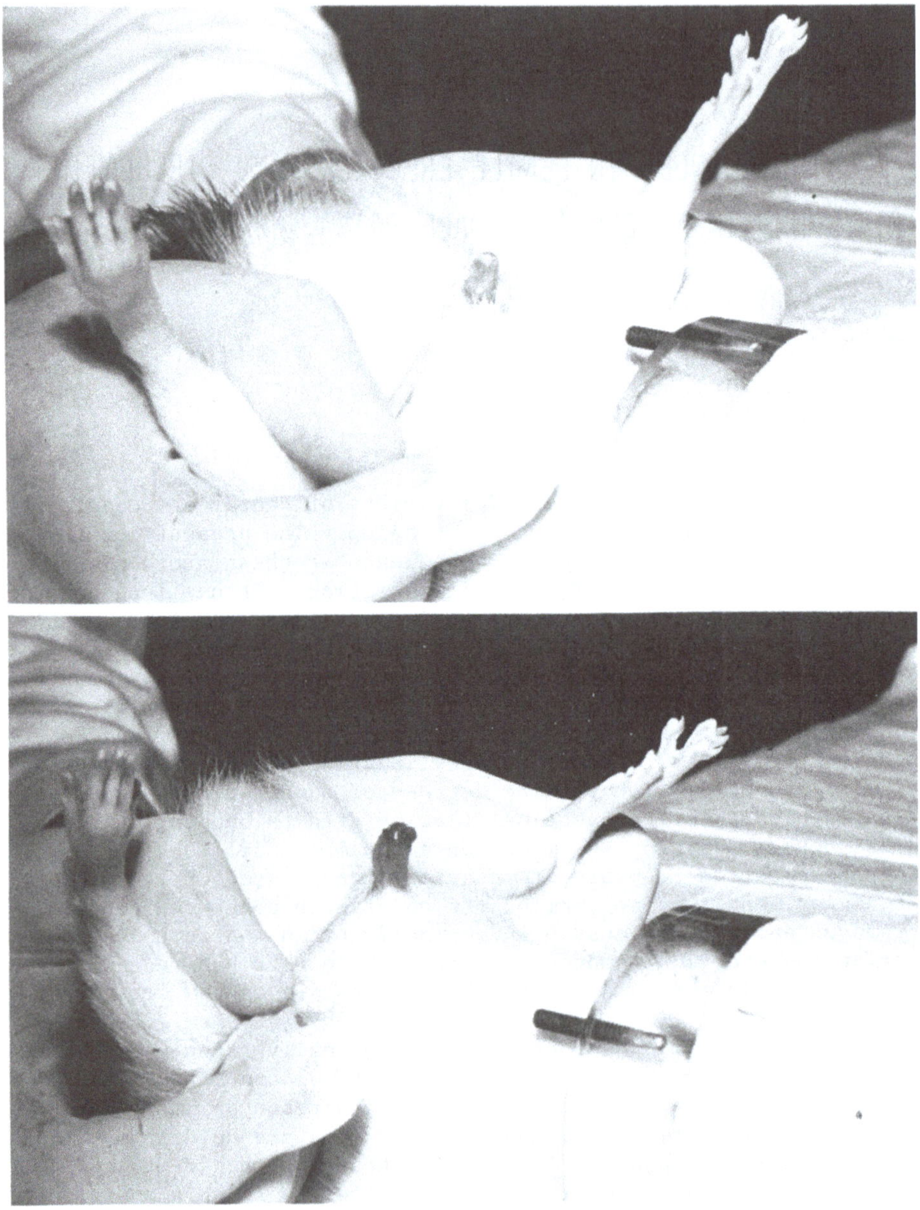

Figure 1. *Testing procedure for the erection response in male rats. The top panel shows a male rat being held on its back, with its penis extruded by retracting the penile sheath; there is not erection. The bottom panel shows a male rat with a fully erect penis (note the slight flaring of the tip).*

(cupping). In normal mating, erections with cupping are associated with ejaculation, but in tests of penile erections, ejaculation is observed only rarely. The erection responses tend to occur in groups or clusters (2-5 erections spaced 5-10 seconds apart), and these clusters are separated by approximately 1-2 minutes. These patterned, phasic responses may continue for over 30 minutes, with no apparent stimulus other than continuous extrusion of the penis from the sheath. No other overt behavior is evidenced by the animal during the erection response. In our testing procedure, aniare observed for 20 minutes after the occurence of the first erection; if no erection occurs within the first 15 minutes of testing, the test is discontinued and scored as negative. The above behavioral technique provides a discrete, quantitative measure of penile erections in male rats.

Examination of age-related changes in the erection response was conducted in middle-aged rats, 13-15 months of age. This age was selected because it is during this period that Long-Evans rats begin to show a significant reduction in circulating levels of luteinizing hormone (LH) and testosterone (Figure 2). These endocrine effects occur in apparently healthy animals in the absence of any general pathological condition, and thus they can be considered as general characteristics of aging in Long-Evans rats *(Gray in press)*. Examination of erection responses during the initial period of the hormonal change should provide an excellent opportunity to establish clear-cut functional relationships between endocrine and behavioral phenomena in the aging process.

Two groups of middle-aged rats were examined: sexually experienced and sexually naive. The sexually-experienced rats had been mated with females and tested for penile erections at three months of age; they were then housed in single-sex, group cages until middle-age. The sexually-naive rats had been maintained in single-sex, group cages from the time of weaning. The two middle-aged groups were compared to a group of young (3 months) sexually-naive rats.

Animals in both middle-aged groups demonstrated a substantial decline in erections responses. They had significantly fewer positive tests than the young animals, and in tests which were positive, they showed significantly fewer erection responses (Figure 3). There was also a tendency for the middle-aged rats to have a longer latency to first erection and a shorter erection duration, but because of the limited number of positive tests for middle-aged animals, the difference was not statistically significant. Sexual experience made no difference in the behavior of the middle-aged rats; the sexually-experienced and sexually-naive groups did not differ significantly in either the percent of positive tests or the total number of erections. Circulating testosterone levels were also evaluated in the three groups of rats. Blood samples were obtained from each animal one week before the behavioral tests, and plasma testosterone was measured by radioimmunoassay. Both groups of middle-aged rats had significantly lower plasma testosterone than the

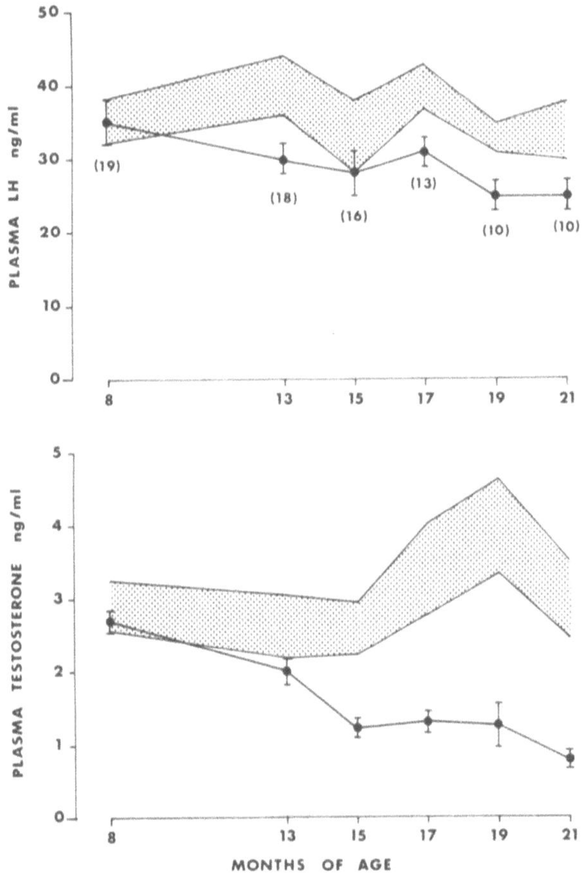

Figure 2. Plasma levels (means ± SE) of LH and testosterone in a group of aging male rats (solid line) and control groups of young 3-4 months male rats (shaded area). The number of healthy aging rats included at each age is given in parentheses. (From Gray, in press).

young rats (young = 3.4 ± 1.0 ng/ml; middle-aged, sexually-experienced = 1.3 ± 0.6; middle-aged, sexually-naive = 1.6 ± 0.8). Since the erection response in male rats is androgen-dependent *(Hart, 1967; Davidson et al., in press)*, the correlation between age-related changes in penile erections and circulating testosterone obviously suggests that the endocrine change may be the causative factor in the behavioral decline.

The possibility of such a functional relationship was investigated by directly manipulatiing circulating testosterone levels in young and middle-aged rats and then comparing erection responses

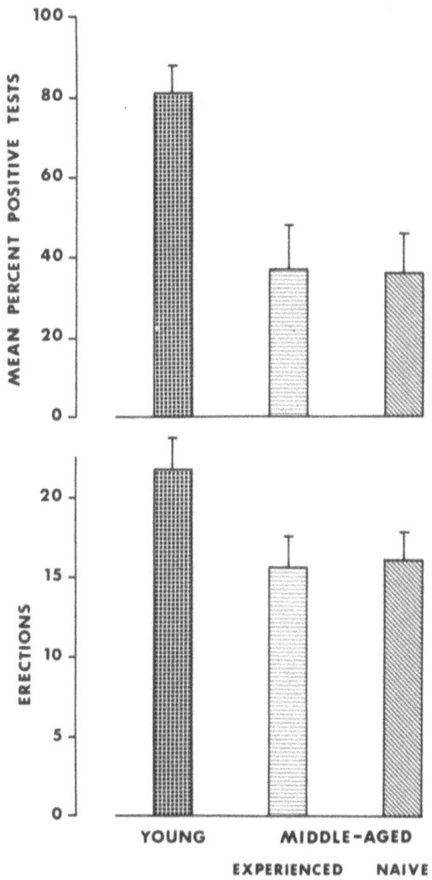

Figure 3. *Erection responses in intact middle-aged (sexually-experienced and sexually-naive) and young (sexually-naive) male rats. Each animal received a series of tests and was evaluated on: (1) the percent of tests on which at one erection occurred; and (2) the mean number of erections in all positive tests. The top panel shows the overall mean (± SE) percent positive tests; the bottom panel shows the mean (± SE) number of erections per postive test. The number of animals in each group was: Young = 15; Middle-aged sexually-experienced = 13; Middle-aged Sexually-Naive = 13.*

in the two age groups. If testosterone changes are the causative factor in the behavioral decline, then equating testosterone levels in the two age groups should eliminate the age differences in erection responses. The procedure involved castrating the animals and implanting them with testosterone-filled Silastic capsules *(Smith,*

Damassa and Davidson, 1977). The length of the capsules was adjusted to produce equal circulating levels of testosterone in both young and middle-aged rats. The levels were validated by blood sampling and radioimmunoassay of plasma testosterone.

In one experiment, supraphysiological levels of testosterone were administered to middle-aged and young rats. They were implanted with Silastic capsules which produced testosterone levels of approximately 7 ng/ml, twice the mean level of intact young rats. Erection tests were conducted three to five weeks later. The middle-aged rats were not significantly different from the young rats in the number of erections (Figure 4). Moreover, their erection frequency was comparable to that seen in young intact rats. Thus, the reduction in the erection response of middle-aged rats is not irreversible. Testosterone administration can eliminate the age-related changes. These results support the view that the decline in the erection response of middle-aged rats is due to decreased levels of circulating testosterone. There is, however, an alternative possibility. The behavioral decline may result from changes

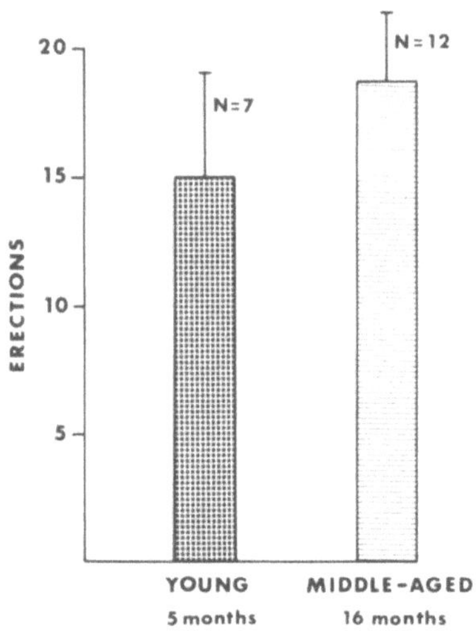

Figure 4. Erection responses in castrated middle-aged and young rats with supraphysiological levels of testosterone. The results represent the mean (± SE) number of erections in all tests, including the negative tests. N = number of animals per group.

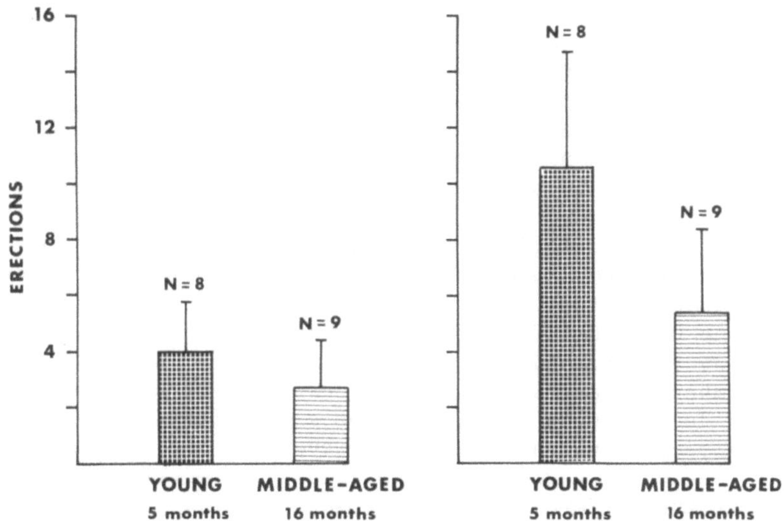

Figure 5. Erection responses in castrated middle-aged and young rats with subphysiological levels of testosterone. The results represent the mean (± SE) number of erections in all tests, including the negative tests. The left panel illustrates the results with 0.5 ng/ml plasma testosterone; the right panel illustrates the results with 1.0 ng/ml plasma testosterone. N = number of animals per group.

in the neural or somatic mechanisms regulating erections, changes which would be reflected in a reduced sensitivity to testosterone. The absence of any age-related differences in animals with supraphysiological testosterone levels may be the result of a "ceiling effect." Young rats may have been responding maximally before castration, with little possibility for an increase. In contrast, the middle-aged rats may have been able to show a substantial increase because of their initially low response rate.

In order to investigate possible age differences in testosterone sensitivity, middle-aged and young animals were administered low levels of testosterone around the threshold for maintaining the behavior. The animals were first implanted with Silasic capsules which yielded testosterone levels of 0.5 ng/ml and tested for erections three weeks later. They were then re-implanted with somewhat larger capsules which produced levels of 1.0 ng/ml and again tested after three weeks. The number of erections elicited from both middle-aged and young rats was low, especially at 0.5 ng/ml testosterone (Figure 5). Apparently, these particular testosterone levels are in fact around the threshold for maintaining the erection response. The middle-aged animals were not significantly different from the young animals in the number of erections at either testosterone level although their mean was lower in both cases. This lack of any statistically significant evidence for differential sensitivity would suggest that there are no age-related changes in the neural and somatic mechanisms regulating erection. This conclusion must be considered tentative at present since the possibility of some age difference in testosterone sensitivity cannot be excluded. The number of animals included in the above experiments is relatively small for behavioral-aging phenomena and only a limited aspect of the dose-response range was investigated. Moreover, the tendency of middle-aged rats to show lower sensitivity at 0.5 and 1.0 ng/ml testosterone is suggestive of some age-related decrement. Additional experimentation is needed with increased numbers of animals and a broader range of hormone levels to establish definitively the absence of any differential sensitivity in middle-aged and young rats. However, if there is in fact an age difference in testosterone sensitivity, the present results indicate that the difference cannot be substantial. It is also unlikely to account entirely for the middle-aged decline in the erection response. The reduction in circulating testosterone must, therefore, be considered a primary factor in the behavioral decline.

This conclusion regarding the relationship between endocrine and behavioral changes in the aging process cannot be extended to all aspects of male rat sexual behavior. The erection response as measured experimentally is a relatively simple, discrete response. Other components of sexual behavior such as the basic pattern of copulatory behavior and sexual arousal are more complex, and recent research in this laboratory indicates that these behaviors demonstrate different age-related changes. The copulatory behavior pattern (i.e., number of intromissions to ejaculation, the time required for ejaculation) does not differ significantly in middle-age and young animals. On the other hand sexual arousal or motivation (what might be described as the tendency to approach the female and initiate copulation) is lower in middle-aged animals. Prior sexual experience is very important in this behavioral effect, unlike the decline in erection responses, and the middle-aged change is much more apparent in sexually-naive animals. Sexual arousal also differs from the erection response in that the age-related change is apparently not eliminated by testosterone administration.

There is the obvious inclination to generalize these results to humans. The neuroendocrine basis for sexual behavior in men is well established,and there are broad similarities between humans and rats *(Davidson, 1977)*. More specifically, a quantitative relationship between circulation testosterone and the erection response have recently been demonstrated in hypogonadal men receiving various doses of testosterone enanthate *(Davidson, in press)*. Human males also show a decline in both circulating free testosterone and the erection response with increasing age *(Kinsey et al.,*

1948; Pirke and Doerr, 1970; Stearns, MacDonnell, Kaufman, Padua, Lucman, Winter and Faiman, 1974). There is then a reasonable basis for assuming a relationship between the age-related changes in testosterone and the erection response. Yet, there are no data which clearly establish such a relationship. The paucity of systematic, quantitative data is surprising, especially in light of the plethora of conflicting, undocumented reports of "androgenic" treatment for impotency symptoms in aging males. One must recognize that penile erection in men is a complex response, much more so than the experimentally-induced erection response in rats. A variety of factors, including non-physiological ones, may be involved in any age-related changes in the erection response in men. There is also the problem of developing valid behavioral tehcniques for measuring the response in humans and separating it from other, even more complex aspects of human sexuality. There is the obvious need for careful, quantitative studies on specific aspects of human sexuality and the physiological basis for any age-related changes. Data in this area on male rats should at the very least provide definitive hypotheses on which to base experimental and clinical studies in humans. The animal data are also of value in a broader sense. It has been suggested that neuroendocrine changes are critical factors in aging *(Finch, 1976).* Male rat sexual behavior may prove a useful experimental model for analyzing general neuroendocrine-behavior relationships in the aging process.

SUMMARY

The relationship between aging changes in pituitary-testicular function and the penile erection reflex was examined in middle-aged male rats. Significant decline in circulating LH and testosterone is first apparent in male rats during middle-age (13-15 mos). In Experiment 1, middle-aged (13-15 mos) and young (3-4 mos) rats were tested for erection responses. The erections were measured in animals held supine and mechanically stimulated by retraction of the penile sheath. Middle-aged rats had substantially fewer erections than young animals. In Experiment 2, groups of middle-aged and young animals were castrated and provided equivalent levels of circulating testosterone (T) through the use of Silastic capsules. The levels were measured directly by radioimmunoassay of plasma T. In animals with supraphysiological T levels, erection frequency was low, but there was no significant difference between middle-aged and young animals. The results suggest that middle-aged rats do not have irreversible deficiencies in the somatic and neural mechanisms regulating erection nor are they significantly less responsive to androgen stimulation. The decline in circulating testosterone with age may therefore be a primary factor in the reduction of penile erections in middle-aged rats.

ACKNOWLEDGEMENTS

The author was supported by a National Research Service Award from the National Institute for Aging. The research was funded by NIH grants MH 21178 and HD 00778 to J.M. Davidson. The assistance of Dr. E.R. Smith, D. Tallentire, C. Spencer and B. Siddal is gratefully acknowledged. Radioimmunoassay reagents were kindly supplied by Drs. G.D. Niswender, L.E. Reichert, B. Caldwell and the NIAMDD pituitary hormone distribution program.

REFERENCES

Davidson, J.M. (1977). Neurohormonal bases of male sexual behavior. In: *Reproductive Physiology II, International Review of Physiology, Vol. 13*, pp. 225-254. Ed. R.O. Greep. Baltimore: University Park Press.

Davidson, J.M. Gonadal hormones and human behavior. In: *Hormonal Contraceptives, Estrogen and Human Welfare.* Eds. M.C. Diamond and C.C. Korenbrot. New York: Academic Press. (in press)

Davidson, J.M., Stefanick, M.L., Sachs, B.D. and Smith, E.R. Role of androgen in sexual reflexes of the male rat. *Physiol. and Behav.* (in press)

Finch, C.E. (1976). The regulation of physiological changes during mammalian aging. *Q. Rev. Biol. 51*, 49-83.

Gray, G.D. Changes in the levels of luteinizing hormone and testosterone in the circulation of aging male rats. *J. Endocr.* (in press)

Hart, B.L. (1967). Testosterone regulation of sexual reflexes in spinal male rats. *Science 155*, 1282-1284.

Hart, B.L. (1968). Sexual reflexes and mating behavior in the male rat. *J. comp. physiol. Psychol. 65*, 453-460.

Kinsey, A.C., Pomeroy, W.B. and Martin, C.E. (1948). *Sexual Behavior in the Human Male.* Philadelphia: Saunders.

Pirke, K.M. and Doerr, P. (1970). Age-related changes in free plasma testosterone, dihydrotestosterone and oestradiol. *Acta Endocr. 80*, 171-178.

Rodgers, C. and Alheid, G. (1972). Relationship of sexual behavior and castration to tumescence in the male rat. *Physiol. and Behav. 9*, 581-584.

Sachs, B.D. and Garinello, L.D. Interaction between penile reflexes and copulation in male rats. *J. comp. physiol. Psychol.* (in press).

Smith, E.R., Damassa, D.A. and Davidson, J.M. (1977). Hormone administration: peripheral and intracranial implants. In: *Methods in Psychobiology, Vol. III*, pp. 259-279. Ed. R.D. Myers. New York: Academic Press.

Stearns, E.L., MacDonnell, J.A., Kaufman, B.J., Padua, R., Lucman, T.S., Winter, J.S.D. and Faiman, C. (1974). Declining testicular function with age. *Am. J. Med. 57*, 761-766.

AGE EFFECTS ON THE HYPOTHALAMIC-PITUITARY-GONADAL CONTROL SYSTEM IN THE RAT

G. D. RIEGLE and A. E. MILLER

Endocrine Research Unit
Department of Physiology
Michigan State University
East Lansing, Michigan 48824

INTRODUCTION

Early recognition that hormones were major regulators of body function has made studies of aging on endocrine control mechanisms attractive to gerontologists. Age alterations in reproductive control systems are well documented. A better understanding of aging changes in the reproductive control system of both man and other species should not only aid in the identification of appropriate animal models to use in the study of aging effects on human physiological systems, but also allow development of clinical procedures which could restore certain endocrine functions and improve the quality of life with advancing age.

In the past few years there has been considerable discussion among gerontologists studying reproductive function whether the age-related deterioration of reproduction is due to alterations in gonadal, pituitary, or hypothalami-neural tissues. Our approach to this problem has been to consider the effect of aging on gonadal, pituitary, and hypothalamic components of this control system in male and female Long-Evans rats. Our rats are maintained in standard colony conditions of controlled temperature and lighting with Wayne Lab-Blox and water supplied *ad libitum*. In these conditions male rats have a 50% survival rate at 21 to 24 months and females from 22 to 26 months with maximal survival of both sexes of about 36 months.

This presentation reviews some recent studies from our laboratory and others on the effect of aging on reproductive control systems.

AGING EFFECTS ON REPRODUCTIVE FUNCTION IN THE MALE

Mammalian aging is characterized by a progressive decrease in sexual activity. Bishop *(1970)* reported progressive testicular atrophy and degeneration in aging males. However, it is clear that there are individual differences in age effects on testicular function. Several reports *(Biship, 1970; Seymour, Duffy and Koerner, 1935)* have shown individual males to be fertile at advanced ages.

There has been considerable disagreement on the effect of age on testosterone in the human male. Kent and Acone *(1966)*, Gandy and Peterson *(1968)* and others have reported no change in blood testosterone in aged men. However, more recent studies *(Baker, Burger, Kekretser, Hudson, O'Conner, Wang, Mirovics, Court, Dunlop and Rennie, 1976; Vermeulen, 1976)* have shown decreased serum testosterone with age. Vermeulen found the decrease in serum testosterone was accompanied by an increase in plasma testosterone binding capacity, which resulted in a further reduction in free testosterone in the older man.

There have been fewer studies of the effects of age on testosterone in laboratory species. Studies by Eleftheriou and Lucas *(1974)* and Finch *(1978)* showed no change in blood testosterone or testicular response to LH in aged mice. On the other hand Leathem and Albrecht *(1974)* reported decreased testicular Δ^5-3β-hydroxysteroid dehydrogenase activity in aged male Long-Evans rats.

Our recent studies on testicular secretion *(Miller and Riegle, 1978)* have shown sharply reduced testosterone concentrations in 22- to 24-month old male rats compared to 4-month old young adults (Figure 1). In this study both the young and aged groups received intravenous injections of saline or saline containing 1, 5, and 20 I.U. of human chorionic gonadotropin (HCG). Testosterone was measured in serial blood samples taken before HCG injection and at 45, 90, and 150 minutes after intravenous injection. HCG stimulation increased serum testosterone concentrations in old groups. The increase in testosterone was greater in the young than in the aged group for each level of HCG tested. These data suggest that although the aged male rat has decreased serum testosterone and reduced responsiveness to acute HCG stimulation, the testis is responsive to stimulation and may be capable of sustaining higher blood testosterone concentrations.

It is accepted that testicular testosterone secretion occurs in episodical bursts rather than continuous sustained secretion. This episodical secretion induces a substantial range of serum hormone concentrations in normal males. In another study, we tested the effect of age on the diurnal pattern of testosterone secretion. In this study, groups of 14 young (4-month) medium (13-month) and aged (22-month) male rats were bled at 2-hour increments that collectively constituted a 24-hour pattern. Serial blood samples were randomly scheduled with at least a 36-hour

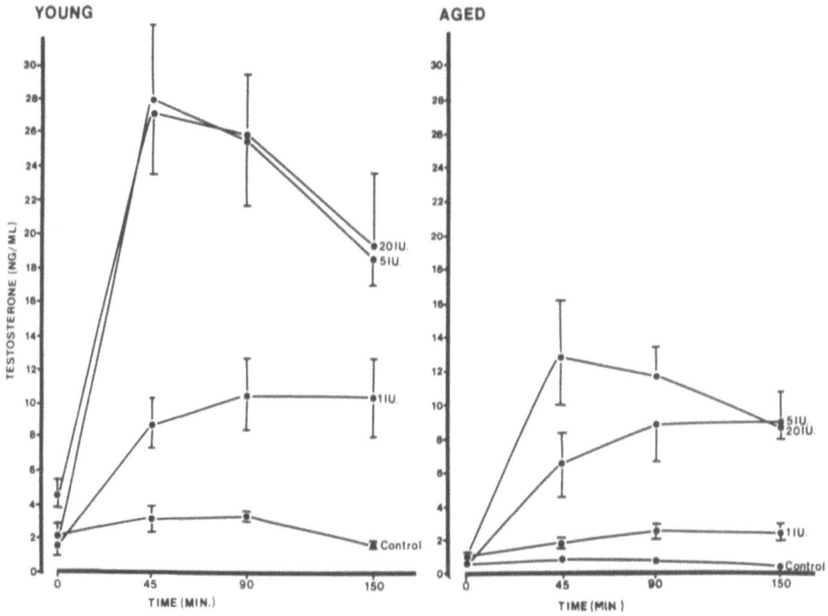

Figure 1. Effect of intravenous injection of 1, 5, and 20 IU of HCG on serum testosterone in young and aged male rats. Testosterone levels are shown as the group mean with indicated SEM from blood samples taken under light ether anesthesia before and at 45, 90, and 150 min after HCG injection.

interval between bleedings to eliminate stress inhibition of the gonadal control system. The results of this study are plotted in Figure 2. Serum testosterone concentrations were consistently higher in young than the aged groups. Both the young and medium aged groups showed a diurnal pattern of testosterone concentration which was absent in the aged group. In addition the aged male showed much less variablilty in mean testosterone levels than did the younger groups. The overall average of testosterone in the 12 blood samples was 4.06 ng/ml in the young group, 1.91 ng/ml in the medium groups and 1.26 ng/ml in the aged group. The range of testosterone concentrations in the 12 serial samples from each rat averaged between 1.02 and 9.67 ng/ml in the young group, between 0.78 and 4.44 ng/ml in the medium group, and between 0.66 and 2.64 ng/ml in the aged male group. The results of this experiment confirm our finding that the aged male rat has reduced serum testosterone and suggests that the episodical testosterone ranges are also reduced in the aged rat.

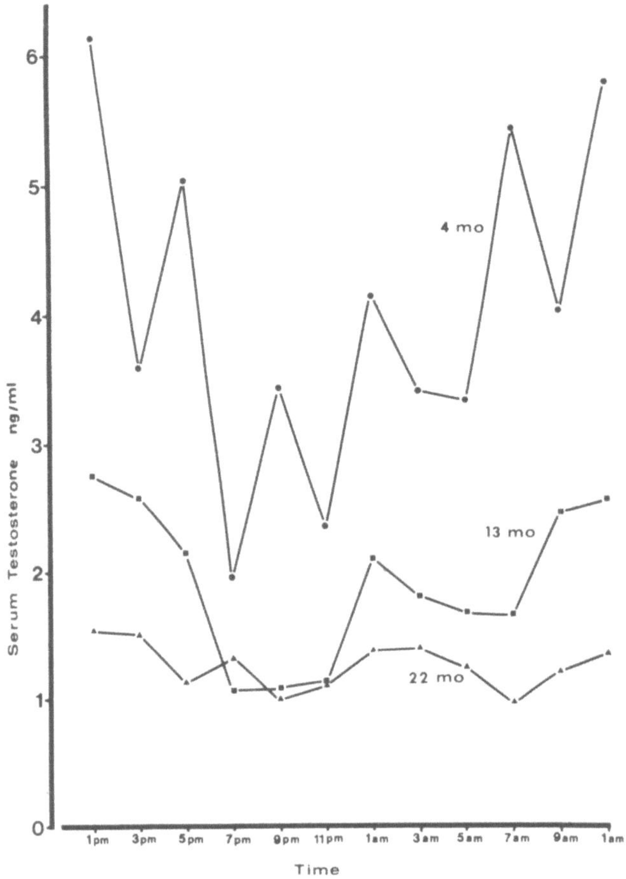

Figure 2. Temporal changes in serum testosterone in young, medium and aged male rats. Testosterone is plotted as group means as a function of the time of sampling. Serial blood samples were taken from each rat in a random sequence under light ether anesthesia. Repeat samples were scheduled at least 18 hours apart.

Leathem and Albrecht *(1974)* showed that they could restore Δ^5-3β-hydroxysteroid dehydrogenase in aged male rats with 5 days of HCG treatment. Following their lead, we tested the effect of 7 days of subcutaneous HCG injection (5 IU HCG/100g body weight for 7 days) on resting testosterone concentration and testicular response to intravenous HCG stimulation *(Miller and Riegle, 1978)*. This study showed that resting testosterone concentrations and testicular response to HCG stimulation were similarly stimulated in young and aged groups following the chronic HCG treatment (Figure 3). These data confirmed our earlier hypothesis that the aged testis was capable of sustaining greater testosterone secretion

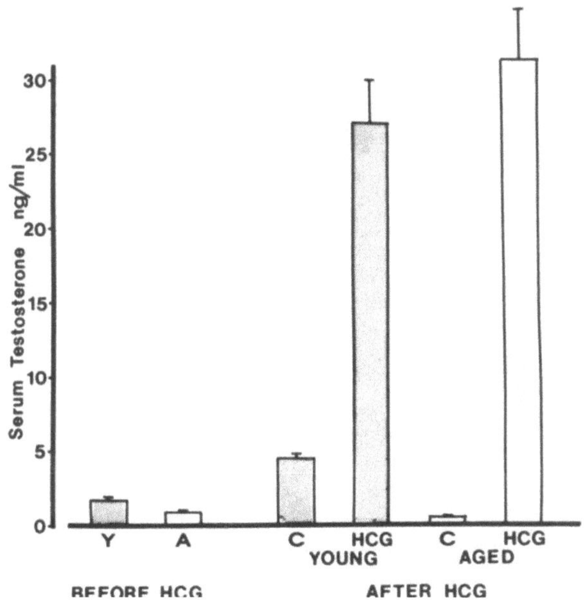

Figure 3. *Effect of daily subcutaneous injection of HCG (5 IU HCG/100 g) for 7 days on serum testosterone levels in young (shaded bars) and aged (open bars) male rats. Blood samples were taken before HCG treatment was started and 24 hours after the final HCG injection. Young and aged control groups (C) received seven treatments of the injection vehicle. Testosterone levels are shown as the group mean with indicated SE.*

and suggest that the testis is not the primary tissue responsible for the aging decline in reproductive function. The experiments also suggest that the primary cause of reduced gonadal endocrine function with age in the male rat is inadequate gonadotropin stimulation of the testis.

The most consistent effect of aging on gonadotropin secretion is the increase in these hormones in the postmenopausal human female *(Heller and Heller, 1939)*. The effect of age on gonadotropin secretion in the male has been more controversial. Vermeulen *(1976)* and Baker *et al., (1976)* found somewhat higher levels of plasma LH and FSH in aged human males. In contrast, there is no evidence for increased gonadotropin secretion accompanying decreased reproductive function in the aged rat. We have consistently found decreased serum LH levels in male rats over 20-months of age *(Riegle and Meites, 1976; Shaar, Euker, Riegle and Meites, 1975)*.

In several experiments, we have tested the responsiveness of the aged rat pituitary to LHRH stimulation. The increase in serum LH following a single LHRH injection is less in aged (24-month) compared

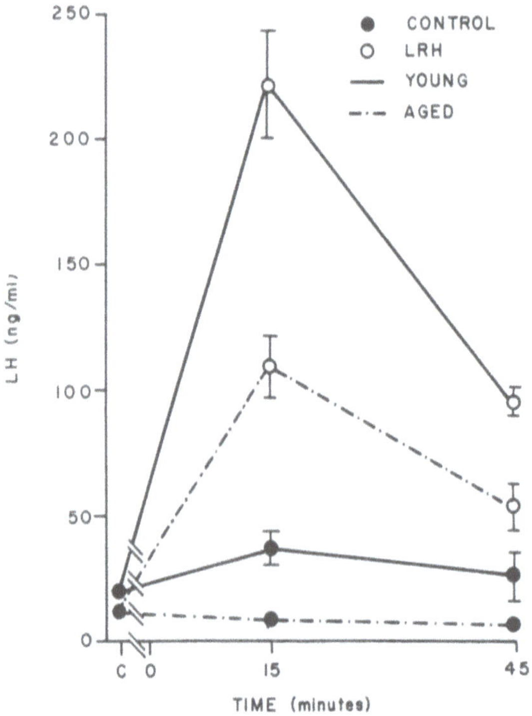

Figure 4. Effect of intravenous injection of 500 ng of LHRH on serum LH in young and aged male rats. Serum LH is shown as the group mean with indicated SEM from serial blood samples taken before and at 15 and 45 minutes after HCG injection.

to young (4-month) male rats (Figure 4). We also found reduced LH release in aged male rat pituitaries incubated in m 199 with 25 to 100 ng of LHRH compared to similarly tested young male rat pituitaries *(Riegle, Meites, Miller and Wood, 1977)*. On the other hand, serum LH was similarly increased in young and aged groups of male rats given multiple LHRH stimulations *(Miller and Riegle, 1978)*. In addition, similar pituitary responsiveness to GNRH has been reported for young and 28 month old C57BL/6J mice *(Finch, Jonec, Wisner, Sinha, de Vellis and Sverdloff, 1978)*.

Our work on gonadotropin secretion in the aged rat clearly

shows reduced serum LH levels and reduced pituitary responsiveness to acute LHRH stimulation. However,the aged rat pituitary consistently responds to LHRH stimulation,and the aged rat can sustain similar serum LH concentrations when the animal receives more sustained stimulation. These data indicate that the pituitary is not the primary tissue responsible for reduced reproductive function in the aged rat and suggest that inadequate hypothalamic stimulation may be involved with aging effects on both the pituitary and gonads.

It is accepted that the hypothalamus synthesizes and secretes hormonal factors that are major regulators of anterior pituitary secretion. Current concepts suggest that hypothalamic hormones are synthesized in discrete regions of the hypothalamus and transported to the medium eminence where they are stored in axon terminals. A massive amount of experimental evidence suggests that the release of stored hormones from axon terminals involves the neurotransmitters of the central nervous system, particularly the catecholamines *(Fernstrom and Wurtman, 1977)*. Fuxe and Hokfelt *(1969)* demonstrated the presence of dopamine containing neurons which terminated in the external layer of the medium eminence, and showed that their dopamine content varied with alterations in pituitary gonadotropin secretion. Kamber, Schneider and McCann *(1970)* showed that intraventricular injection of dopamine stimulated pituitary gonadotropin secretion and inhibited prolactin secretion. On the other hand, others have reported that the release of hypothalamic gonadotropin releasing hormone could be stimulated by norepinephrine *(Cocchi, Fraschini, Jolanbo, and Mullers, 1974; Sawyer, Hilliard, Kanematsu, Scaramuzzi and Blake, 1974)*.

We and others have hypothesized that the hypothalamus is a primary site of age-related alterations in reproductive control systems. We have considered age effects of hypothalamic function in terms of tissue content of gonadotropic releasing hormone, catecholamine content, and hypothalamic responsiveness to stimulatory and inhibitory input.

In a series of experiments,we compared hypothalamic LH-releasing activity in hypothalamic extracts from young and aged male and female rats. The hypothalamic island was removed and homogenized in cold 0.4 N perchloric acid. After centrifugation,the extract was adjusted to a pH of 7.25 and aliquots of 0,0.25,0.50 or 1.0 hypothalamic equivalents were added to anterior pituitary incubates from young donor rats. The results of these experiments (Figures 5 and 6) showed that hypothalamic extracts stimulated LH release from the incubated rat pituitaries. In addition, these studies showed similar LH releasing activities in hypothalamic extracts from gonadectomized and intact young and aged male and female rats. These results indicate that aged rat hypothalami from both of the physiological states tested, aged intact rats which are secreting low levels of LH and aged gonadectomized rats with stimulated LH secretion, have similar amounts of LH simulating activity as was found in similar young groups.

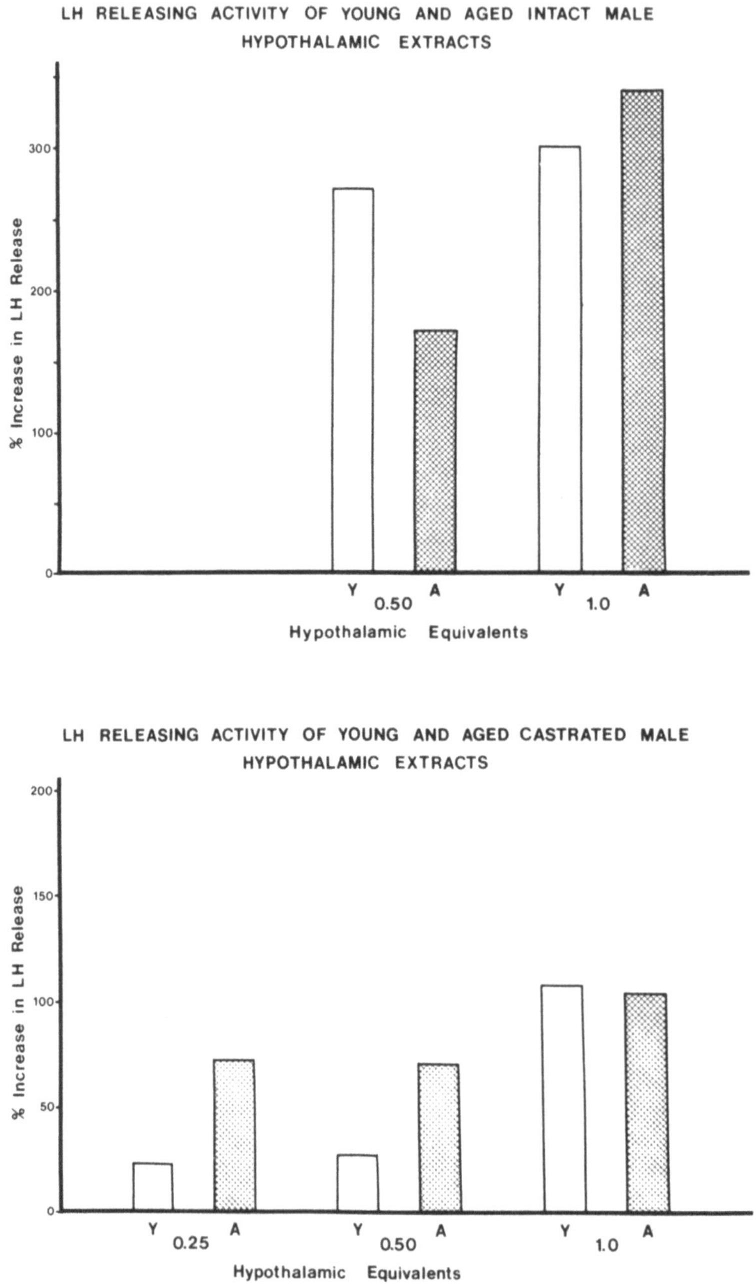

Figure 5. Effect of 0, 0.25, 0.50, and 1.00 hypothalamic extract equivalents from young (Y) and aged (A) male rats on LH release from incubated rat pituitaries. LH release is shown as percent increase over LH release from similarly incubated paired control pituitary halves.

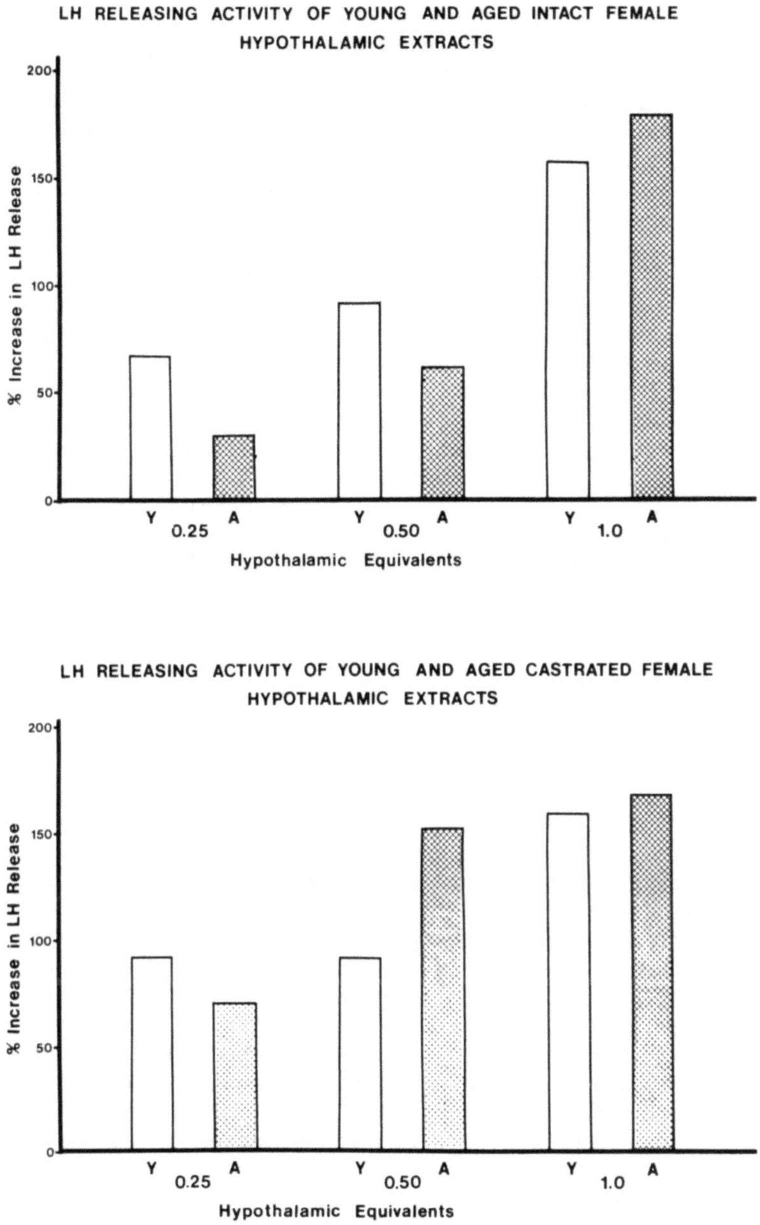

Figure 6. Effect of 0, 0.25, 0.50, and 1.00 hypothalamic extract equivalents from young (Y) and aged (A) female rats on LH release from incubated rat pituitaries. LH release is shown as percent increase over LH release from similarly incubated paired control pituitary halves.

This finding suggests that the decreased LH secretion of the aged rats may reflect failure to release the hypothalamic hormone rather than a decrease in neuronal content of LHRH.

Palkovits, Brownstein, Saavedra and Axelrod *(1974)* showed substantial hypothalamic content of norepinephrine and dopamine. There is growing evidence that these hypothalamic monoamines are important in the regulation of gonadotropin secretion. We measured hypothalamic dopamine and norepinephrine content by microfluorescence of acid extracts of hypothalamic tissue *(Miller and Riegle, 1976)*. We found significant reductions in hypothalamic content of both monoamines in aged (24- to 26-month) compared to young (4-month) male rats. Dopamine content was reduced from 32.5 ng in young rats to 15.6 ng/hypothalamus in the aged rats. Norepinephrine was similarly reduced from 47.6 ng in young to 22.8 ng/hypothalamus in the aged male rats. These findings have been confirmed by more recent studies by Finch *(1978)* and Meites, Huang, and Simpkins *(1978)* and suggest that alterations in these neurotransmitters are involved with the loss of reproductive function in the aged laboratory rodent.

In a series of experiments we have tested the responsiveness of the hypothalamic-pituitary reproductive control unit to stimulatory and inhibitory input. In one study,we considered the effects of gonadectomy and administration of testosterone on serum LH in young and aged male rats *(Shaar, et al., 1975)*. Serum LH levels increased more rapidly after gonadectomy in the young than in the aged groups and remained higher throughout the experiment. Serum LH levels in gonadectomized young were about twice that of the aged groups at the end of the experiment. On the other hand, the aged, gonadectomized rats were more responsive to testosterone inhibition of serum LH than the young rats suggesting that the hypothalamic-pituitary unit is more sensitive to steroid negative feedback in the aged male rat.

In an earlier experiment, we found increased serum LH in young male rats exposed to stress *(Euker, Meites and Riegle, 1975)*. Although acute stress results in a prompt increase in serum LH in young male rats, the treatment was without effect in aged male rats *(Riegle and Meites, 1976)*, supporting the hypothesis of decreased responsiveness of the gonadotropin control mechanism in the aged male rat.

L-dopa administration is presumed to increase neuronal catecholamine function. Lu and Meites *(1972)* showed that systemic L-dopa injections could lower serum prolactin in rats. We found decreased responsiveness of aged male rats to L-dopa inhibition of prolactin release compared to similarly treated young groups *(Riegle and Meites, 1976)*.

In summary, these studies indicate that there are major changes in hypothalamic control of pituitary secretion in the aged male rat that result in decreased serum LH and testosterone concentrations. The hypothalamus of the old rat contains sufficient releasing hormones to affect pituitary secretions. The primary

TABLE 1. THE EFFECT OF AGE ON REPRODUCTION IN THE RAT

Age (months)	Litter no.	No. rats	No. mated	No. littered	Average no.pups	Average pup wt. (g)	Proestrus LH (ng/ml)
2	1	12	12	10	10.3	5.7	796
4	2	12	12	11	8.2	6.1	625
6	3	12	12	8	9.1	6.2	1306
7.5	4	12	10	8	7.4	7.4	898
9	5	24	18	14	9.1	6.0	922
11	6	24	20	11	9.5	6.0	687
13	7	23	17	6	6.9	6.4	851
14.5	8	23	17	3	7.0	7.0	691
17	9	22	13	1	9.0	5.0	925

change occurring with age involves altered sensitivity to control input. There is substantial evidence implicating age changes in hypothalamic monoamines with the loss of reproductive function.

AGING EFFECTS ON REPRODUCTIVE FUNCTION IN THE FEMALE

The best known aging effect on female reproduction is the loss of ovarian function that occurs in women. Although the number of oocytes is exhausted soon after menopause in women, other species experience reproductive failure with substantial numbers of oocytes remaining in the ovaries *(Talbert, 1978)*. In most rodent species,the loss of oocytes with increasing age is constant and rarely reaches zero before death. Adams *(1970)* showed a more rapid decline in litter size than in ovulation rate. This finding is supported by Thorneycroft and Soderwall *(1969)* who found increased post-ovulatory loss of reproduction in aged hamsters.

We recently studied the effect of age on reproduction in rats subjected to serial matings between 2-months and 17-months of age. Half of the rats were repeatedly mated beginning at 2-months of age with the other half subjected to serial pregnancies beginning at 9-months of age. Parameters of reproductive function measured included the numbers of rats with estrous cycles, numbers mated, numbers littering, mean litter size and weight and serum LH concentrations in a blood sample taken at 4 pm on proestrus. There were no differences in any of these parameters in rats starting this experiment at 2- or 9-months of age.

The results of this experiment are shown in Table 1. The percentages of rats that were having estrous cycles and were successfully mated started to decline at 7.5-months of age. However,60% of the rats were still mated at 17-months of age. The decrease in the number of rats littering declined more sharply with only 1 litter born in the 17-month old rats. There were no differences in litter size or average pup weight with increasing age. This finding differs from the reports of Adams *(1970)* and Talbert *(1978)*. Our rats exhibit an "all or none" response with respect to the loss of fertility. The mated rats seem to either lose their entire litter or deliver normal litters similar to young rats. The cycling rats did not show a decline in their proestrous LH concentration. We have previously shown sharply reduced increases in serum LH following ovariectomy in aged females compared to young gonadectomized groups *(Shaar et al., 1975)*. These findings suggest that the aged female rat is more responsive to the stimulation of the hypothalamus-pituitary unit at proestrus than to the decrease in steroid negative feedback after gonadectomy.

The results of the previous study and work by other investigators suggest that there may be alterations in ovarian progesterone secretion which could contribute to the loss of post-ovulatory reproduction in the aged rodent. The loss of oocytes cause

TABLE 2. THE EFFECT OF AGE ON SERUM PROGESTERONE IN PREGNANT AND PSEUDOPREGNANT RATS

			Progesterone (ng/ml) Days After Mating			
Age (months)	Ovarian state	n	Day 1	Day 6	Day 11	Day 16
4	cycling, littered	12	5.9 ± 0.9	74.0 ± 9.3	63.6 ± 7.2	71.2 ± 6.5
7	cycling, littered	13	3.2 ± 0.4	37.1 ± 2.5	49.0 ± 4.4	64.0 ± 8.1
9	cycling, littered	10	4.4 ± 1.3	43.1 ± 4.3	52.2 ± 5.0	66.5 ± 7.0
	cycling, pseudo	2	2.0	53.7	26.7	16.5
11	cycling, littered	6	1.9 ± 0.4	32.1 ± 5.1	35.1 ± 4.4	35.4 ± 4.7
	cycling, pseudo	5	3.4 ± 0.3	49.8 ± 6.5	37.5 ± 7.1	18.3 ± 6.1
	C.E., pseudo	9	5.2 ± 1.9	54.1 ± 6.3	39.1 ± 4.5	23.9 ± 7.7
13	cycling, littered	3	4.8 ± 2.0	39.3 ± 1.9	43.6 ± 2.0	40.0 ±18.9
	cycling, pseudo	7	3.3 ± 1.3	33.4 ± 3.7	26.6 ± 3.0	20.2 ± 6.9
15	cycling, pseudo	12	4.6 ± 1.0	54.4 ± 7.1	49.4 ± 6.8	32.9 ± 7.9
	C.E., pseudo	9	2.0 ± 0.5	58.5 ± 6.5	37.5 ± 7.1	31.8 ± 7.2
20	cycling, pseudo	8	11.2 ± 3.3	38.3 ± 6.3	44.4 ± 4.2	35.9 ± 8.0
	C.E., pseudo	4	5.9 ± 2.0	38.9 ± 9.6	48.5 ±10.4	11.1 ± 3.8
22	cycling, pseudo	4	3.7 ± 2.8	100.6 ±13.4	87.5 ± 6.8	7.5 ± 2.8
	C.E., pseudo	5	1.8 ± 0.4	87.2 ±19.5	74.8 ± 22.7	22.8 ± 8.9

C.E.: Constant Estrous

ovarian steroid production to decrease dramatically after menopause in women in spite of increased gonadotropin secretion *(Mattingly and Huang, 1969; Procope, 1969; Adamopoulos, Loraine and Dove, 1971)*. Most evidence of aging effects on rat ovarian steroid secretion have been indirect utilizing vaginal cytological changes as indices of sex steroid secretion. Aschheim *(1976)* has characterized aging effects on rat vaginal cytology. Although some rats remain cyclic throughout their lives, increased percentages show constant estrous or repetitive pseudopregnant vaginal smears with increased age. Aschheim reported a predomimance of rats showing constant estrous vaginal smears in the second year of life, with repetitive pseudopregnancies becoming dominant in the third year. Aschheim *(1976)* also found normal numbers of eggs ovulated in aged rats retaining normal estrous cycles.

Conclusive evidence for the endocrine potential of some aged rat ovaries was supplied by Aschheim *(1964/65)*. He found that ovaries grafted from young donors into aged recipients resumed the endocrine state of the recipient while some aged rat ovaries transplanted into young recipients showed normal endocrine cyclicity. Aschheim's results were confirmed by the experiments of Peng and Huang *(1973)*.

We and others have hypothesized that changes in progesterone and estrogen are involved with the loss of reproductive function in aging female rats. The effect of age on serum progesterone during pregnancy and pseudopregnancy was studied after mating in rats with normal ovarian cycles at 4, 7, 9, 11, 13, 15, 20 and 22 months of age. Progesterone was also measured in constant estrous rats at 11, 15, 20 and 22 months of age made pseudopregnant by mating. Progesterone was measured in serial blood samples taken at 4 pm on days 1, 6, 11 and 16 of pregnancy of pseudopregnancy. Successful pregnancy was reduced to 50% in the 11-month and to 25% in the 13-month old groups with no successful pregnancies in the older rats. The effect of age on serum progesterone in shown in Table 2. Progesterone ranged from 1.8 to 11.2 ng/ml on day 1. Progesterone was increased on days 6 and 11 after mating, ranging from 32 to 100 ng/ml. Progesterone concentrations at days 6 and 11 after mating were not different in pregnant or pseudopregnant rats and were not significantly reduced in the age groups. Progesterone averaged 55 ng/ml on day 11 of pregnancy in 43 rats delivering litters and 48 ng/ml on day 11 of pseudopregnancy in 57 rats. Although young rats delivering normal litters maintained high progesterone on day 16 after mating, the pseudopregnant groups had reduced progesterone concentrations. These data suggest that the loss of reproduction in the aged female rat is not due to decreased luteal secretion of progesterone.

Progesterone has an important role in the regulation of the ovarian cycle in the rat. We are currently conducting experiments to measure temporal changes in serum progesterone in young and aging rats during the estrous cycle and in constant estrous and pseudopregnant states. Progesterone has been measured in serial blood samples taken at 4-hour intervals. Young rats show a progressive increase in progesterone during proestrus from 2 ng/ml at 8 AM to 35 ng/ml at 8 PM. Cycling rats at 12 or 20 months of age have much smaller increases in progesterone during proestrus. The aged females also have smaller diurnal surges of serum progesterone in estrus and diestrous stages of the ovarian cycle. Progesterone was higher in aged pseudopregnant and persistent diestrous rats (4-12 ng/ml) than in aged constant estrous groups (1-2 ng/ml). These preliminary data indicate considerable age-related effects on serum progesterone concentration.

Pituitary gonadotropin secretion is markedly altered in aged females. Although Everitt *(1976)* suggests that excessive gonadotropin stimulation of the ovary in aging women could result in exhaustion of the ovary, most investigators have not found increased gonadotropin secretion until after the climacteric *(Coble, Kohler, Cangille and Ross, 1969; Kohler, Ross and Odell, 1968)*. In addition, there is no evidence of increased LH secretion in aged female rats. Aschheim *(1976)* reported the existence of ovarian interstitial deficiency cells in aged rats which could be restored by LH injection. This suggestion of inadequate LH stimulation of aged rat ovaries is supported by our work in the aged female rat which has repeatedly demonstrated decreased serum LH and increased prolactin concentration *(Shaar et al., 1975; Watkins, Meites and Riegle, 1975)*.

We have tested pituitary responsiveness to LHRH in young and aged female rats *(Watkins, et al., 1975)*. Although pituitary LH content and LH secretion after a single LHRH injection are reduced in aged female rats, our work (Figure 7) shows that the aged remale can sustain substantial serum LH concentrations which are similar to that of young females when multiple LHRH injections are made *(Miller and Riegle, 1978)*. These data suggest that the pituitary of the aged female rat can secrete greater amounts of LH than normally occur indicating that the aged rat pituitary does not receive sufficient neuroendocrine stimulation of gonadotropin secretion. This hypothesis is supported by the recent studies of Meites et al. *(1978)*.

There is growing evidence that there are significant alterations in hypothalamic function in the aged female rat which are similar to that previously outlined for the male. Clemens, Amenomori, Jenkins and Meites *(1969)* induced ovulation in aged constant estrous rats by electrical stimulation of the preoptic area. Huang, Marshall and Meites *(1976)* also showed that ovarian cycles could be reinitiated in aged constant estrous rats by stress treatments, or injections of ACTH or progesterone. In addition, Quadri, Kledzik and Meites *(1973)* found that systemic injections of L-dopa or iproniazid, which presumably act by increasing brain catecholamine content could restore vaginal cyclicity in similarly aged constant estrous rats. These experiments suggest that age effects on hypothalamic catecholamine function could be related to the loss of reproduction in the aged

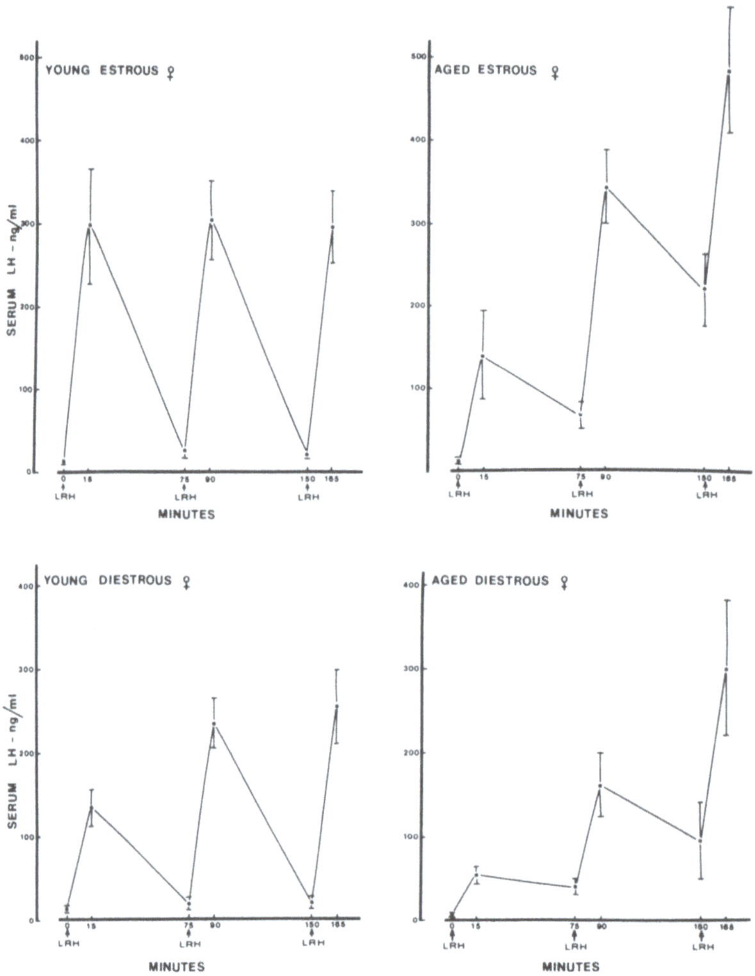

Figure 7. Effect of three intravenous injections of 500 ng of LHRH on serum LH in young estrous and diestrous and aged constant estrous and persistent diestrous female rats. Serum LH is shown as the group mean with indicated SE. Serial blood samples were taken before LHRH injection (0, 75 and 150 min) and 15 min after LHRH stimulation (15, 90 and 165 min).

rat. This hypothesis is directly supported by the work of Clemens and Finch which are reported elsewhere in this volume.

In summary, the effect of age on the responsiveness of hypothalamic-pituitary stimulation of gonadotropin release in unclear. We *(Shaar et al., 1975)* and others have shown significantly smaller increases in serum LH following ovariectomy in aged compared to young female rats. Serum LH concentrations 6 weeks after gonadectomy were about three times greater in young compared to aged female rats. In addition, we found decreased pituitary LH secretion following a single LHRH stimulation *(Watkins et al., 1975)*. However aged rats receiving multiple LHRH stimulations had serum LH concentration that were similar to identically treated young rats *(Miller and Riegle, 1978)*, and our work on the effect of age on reproduction indicated no effect of age on 4 PM proestrous LH concentrations.

CONCLUSION

Experimental evidence reported here indicates that significant changes occur in the gonadotropin control mechanism of the laboratory rat with age. Our work shows that aging is accompanied by decreased serum testosterone, LH, FSH and progesterone and increased serum prolactin. Our work also shows reduced gonadal and pituitary responsiveness to acute stimulation of LH and LHRH. On the other hand, the gonads and pituitaries of aged rats are responsive to stimulation and, under conditions of chronic stimulation, they are capable of secreting similar amounts of hormones as found in similarly treated young rats. Although aged female rats lose their capacity to reproduce by 13 to 17 months of age, our studies suggest that this failure to reproduce is not due to failure of proestrus LH secretion or luteal function. These results indicate that the pituitaries and gonads of aged rats are capable of sustaining greater endocrine function than that normally occurring in these animals.

Our studies suggest that the most fundamental change occurring in the reproductive control system of the aging rat is neuroendocrine, involving hypothalamic regulation of pituitary gonadotropin secretion. We have shown that the hypothalamus contains sufficient LH-releasing activity to affect pituitary secretion. Our experiments indicate major age differences in hypothalamic response to factors that stimulate or inhibit hypothalamic hormone release. A growing body of evidence suggests that there are significant changes in hypothalamic catecholamine content or activity in the aged rat which are related to neuroendocrine function. It is hoped that these findings should contribute to our understanding of the effect of aging on reproduction and permit the development of clinical therapies that can improve the endocrine state of the aged.

REFERENCES

Adamopoulos, D.A., Loraine, J.A. and Dove, G.A. (1971). Endocrinological studies in women approaching the menopause. *J. Obst. Gync. Br. Commonw. 78,* 62-79.

Adams, C.E. (1970). Aging and reproduction in the female mammal with particular reference to the rabbit. *J. Reprod. Fert. (Suppl) 12*, 1-16.

Aschheim, P. (1964/65). Résultats fournis par la greffe hétérochrone des ovaires dan l´ étude de la régulation hypothalamo-hypophyso-ovarienne de la ratte sénile. *Gerontologia 10*, 65-75.

Aschheim, P. (1976). Aging in the hypothalamic-hypophyseal ovarian axis in the rat. In: *Hypothalamus, Pituitary and Aging*, Eds. A.V. Everitt and J.A. Burgess, pp. 376-418. Charles C. Thomas, Springfield, Ill.

Baker, H.W.G., Burger, H.G., deKretser, D.M., Hudson, B., O'Conner, S., Wang, C., Mirovics, A., Cour, J., Dunlop, M. and Rennie, G.C. (1976). Changes in the pituitary-testicular system with age. *Clin. Endocr. (OXF) 5*, 349-372. Association of Gerontology, Kiev.

Biship, M.W.H. (1970). Aging and reproduction in the male. *J. Reprod. Fert. (Suppl) 12*, 65-87.

Clemens, J.A., Amenomori, Y., Jenkins, T. and Meites, J. (1969). Effects of hypothalamic stimulation, hormones, and drugs on ovarian function in old female rats. *Proc. Soc. exp. Biol. Med. 132*, 561-563.

Coble, Y.D., Kohler, P.O., Cargille, C.M. and Ross, G.T. (1969). Production rates and metabolic clearance rates of human follicle-stimulating hormone in premenopausal and postmenopausal women. *J. clin. Invest. 48*, 359-363.

Cocchi, D., Fraschini, F., Jalanbo, H. and Muller, E. (1974). Role of brain catecholamines in the postcastration rise in plasma LH of prepuberal rats. *Endocrinology 95*, 1649-1657.

Eleftheriou, B.E. and Lucas, L.A. (1974). Age-related changes in testes, seminal vesicles and plasma testosterone levels in male mice. *Gerontologia 20*, 231-238.

Euker, J.S., Meites, J. and Riegle, G.D. (1975). Effects of acute stress on serum LH and prolactin in intact, castrate, and dexamethasone-treated male rats. *Endocrinology 96*, 86-92.

Everitt, A.W. (1976). The female climacteric. In: *Hypothalamus, Pituitary and Aging*. Eds. A.V. Everitt and J.A. Burgess, pp. 419-430. Charles C. Thomas, Springfield, Ill.

Fernstrom, J.D. and Wurtman, R.J. (1977). Brain monoamines and reproductive function In: *Reproductive Physiology 11*, Ed. R.O. Greep, International Review of Physiology, Vol. 13. pp. 23-55. University Park Press, Baltimore.

Finch, C.E. (1978). Reproductive senescence in rodents: Factors in the decline of fertility and loss of regular estrous cycles. In: *The Aging Reproductive System*, Ed. E.L. Schneider, pp. 193-212. Raven Press, New York.

Finch, C.E., Jonec, V., Wisner, J.R., Jr., Sinha, Y.N., de Vellis, J.S. and Sverdloff, R.S. (1978). Hormone production by the pituitary and testis of male C57BL/JL mice during aging. *Endocrinology 101*, 1310-1318.

Fuxe, K. and Hikfelt, T. (1969). Catecholamines in the hypothalamus and pituitary gland. In: *Frontiers in Neuroendocrinology* Eds. W.F. Ganong and L. Martini, pp. 47-96. Oxford University Press, Oxford.

Gandy, H.M. and Peterson R.F. (1968). Measurement of testosterone and 17-ketosteroids in plasma by the double isotope dilution derivative technique. *J. clin. Endocr. Metab. 28*, 949-977.

Heller, C.G. and Heller, E.J. (1939). Gonadotropic hormone: Urine assays of normally cycling, menopausal, castrated, and estrin treated human females. *J. clin. Invest. 18*, 171-178.

Huang, H.H., Marshall, S. and Meites, J. (1976). Induction of estrous cycles in old non-cyclic rats by progesterone, ACTH, ether stress, or L-dopa. *Neuroendocrinology, 20*, 21-34.

Kamberi, I.A., Schneider, H.P.G. and McCann, S.M. (1970). Action of dopamine to induce release of FSH-releasing factor (FRF) from hypothalamic tissue *in vitro*. *Endocrinology 86*, 278-284.

Kent, J.R. and Acone, A.B. (1966). Plasma testosterone levels and aging in males. In: *Androgens in Normal and Pathological Conditions*. Eds. A. Vermeulen and D. Exley, pp. 31-35. Excerpta Medica Foundation Intl. Congr. Series 101, Amsterdam.

Kohler, P.I., Ross, G.T. and Odell, W.D. (1968). Metabolic clearance and production rates of human luteinizing hormone in pre- and post-menopausal women. *J. clin. Invest. 47*, 38-47.

Leathem, J.H. and Albrecht, E.D. (1974). Effect of age on testis Δ^5-3β-hydroxysteroid dehydrogenase in the rat. *Proc. Soc. exp. Biol. Med. 145*, 1212-1214.

Lu, K.-H. and Meites, J. (1972). Effects of L-dopa on serum prolactin and PIF in intact and hypophysectomized, pituitary-grafter rats. *Endocrinology 91*, 868-872.

Mattingly, R.F. and Huang, W.Y. (1969). Steroidogenesis of the menopausal and postmenopausal ovary. *Am. J. Obstet. Gynec. 103*, 679-693.

Meites, J., Huang, H.H. and Simpkins, J.W. (1978). Recent studies on neuroendocrine control of reporductive senescence in rats. In: *The Aging Reproductive System*. Ed. E.L. Schneider, pp. 213-235. Raven Press, New York.

Miller, A.E., Sharr, C.J. and Riegle, G.D. (1976). Aging effects on hypothalamic dopamine and norepinephrine content in the male rat. *Exp. Aging Res. 2*, 475-480.

Miller, A.E. and Riegle, G.D. (1978). Serum testosterone and testicular response to HCG in young and aged male rats. *J. Geront. 33*. 197-203.

Miller, A.E. and Riegle, G.E. (1978). Serum LH levels following multiple LHRH injections in aging rats. *Proc. Soc. exp. Biol. Med.*, 494-499.

Palkovits, M., Brownstein, M., Saavedra, J.M. and Axelrod, J. (1974). Norepinephrine and dopamine content of hypothalamic nuclei of the rat. *Brain Res. 77*, 137-149.

Peng, M.T. and Huang, H.H. (1973). Aging of hypothalamic-pituitary-ovarian function in the rat. *Fert. Steril. 23*, 535-542.

Procope, B.J. (1969). Studies on urinary excretion, biological effects and origin of estrogens in post-menopausal women. *Acta Endocr. (Suppl) (Kbh) 135*, 9-110.

Quadri, S.K., Kledzik, G.A. and Meites, J. (1973). Reinitiation of estrous cycles in old constant-estrous rats by central-acting drugs. *Neuroendocrinology 11*, 248-255.

Reigle, G.D. and Meites, J. (1976). Effects of aging on LH and porlactin after LHRH L-dopa, methyl dopa, and stress in the male rat. *Proc. Soc. exp. Biol. Med. 151*, 507-511.

Riegle, G.E., Meites, J., Miller, A.E., and Wood, S.M. (1977). Effect of aging on hypothalamic LH-releasing and prolactin inhibiting activities and pituitary responsiveness to LHRH in the male laboratory rat. *J. Geront. 32*, 13-18.

Sawyer, D.H., Hilliard, J., Kanematsu, S., Scaramuzzi, R. and Blake, C.A. (1974). Effects of intraventricular infusions of norepinephrine and dopamine on LH release and ovulation in the rabbit. *Neuroendocrinology 15*, 328-337.

Schneider, H.P.G. and McCann, S.M. (1970). Release of LH-releasing factor (LRF) into the peripheral circulation of hypophysectomized rats by dopamine and its blockage of estradiol. *Endocrinology 87*, 249-253.

Seymour, R.E., Duffy, C. and Koerner, A. (1935). A case of authenticated fertility in a man of 94. *J. Am. med. Ass. 105*, 1423-1424.

Shaar, C.J., Euker, J.S., Riegle, G.D. and Meites, J. (1975). Effects of castration and gonadal steroids on serum luteinizing hormone and prolactin in old and young rats. *J. Endocr. 66*, 45-51.

Talbert, G.B. (1978). Effect of aging of the ovaries and female gametes on reproductive capacity. In: *The Aging Reproductive System.* Ed. E.L.S. Schneider, pp. 59-83. Raven Press, New York.

Thorneycroft, I.H. and Soderwall, A.L. (1969). Ovarian morphological and functional changes in reproductively senescent hamsters. *Anat. Rec. 165*, 349-354.

Vermeulen, A. (1976). Leydig cell function in old age. In: *Hypothalamus, Pituitary and Aging.* Eds. A.V. Everitt and J.A. Burgess, pp. 458-463. Charles C. Thomas, Springfield, Ill.

Watkins, B.E., Meites, J. and Riegle, G.D. (1975). Age-related changes in pituitary responsiveness to LHRH in the female rat. *Endocrinology 97*, 543-548.

AN ENDOCRINE HYPOTHESIS OF BRAIN AGING AND STUDIES ON BRAIN-ENDOCRINE CORRELATIONS AND MONOSYNAPTIC NEUROPHYSIOLOGY DURING AGING

PHILIP W. LANDFIELD

Department of Physiology and Pharmacology
Bowman Gray School of Medicine
Winston-Salem, North Carolina 27103

As many papers in this symposium indicate, there is rapidly growing evidence that alterations in the neural regulation of endocrine functions may play a major role in mammalian aging. In particular, there seems to be a somewhat selective age-dependent deterioration of dopaminergic transmitter systems in rodents *(Finch, 1973; Simpkins, Mueller, Huang and Meites, 1977)* and in humans *(Carlsson, this volume)*. Additionally, there is now some evidence that reduced dopaminergic presynaptic function is associated with reduced postsynaptic dopamine receptors in at least some systems *(Severinson and Finch, unpublished; Finch, this volume)*. The major role which dopamine seems to play in certain well-established age-dependent syndromes, such as parkinsonism in humans *(Carlsson, this volume; Barbeau, this volume)* or cessation of ovarian cycling in rats *(Clemens, this volume; Clemens and Bennett, 1977)*, as well as the apparent significance of hypothalamic dopaminergic mechanisms in endocrine regulation *(reviewed in Reichlin, 1974; Schally, Arimura and Kastin, 1973; Blackwell and Guillemin, 1973)*, lends significant support to the view that neuroendocrine deregulation could be a critical factor in the mammalian aging process.

Along with an apparently greater susceptibility of dopaminergic systems to aging processes, there is, moreover, evidence that noradrenergic, cholinergic and other putative neurotransmitter systems which probably function in endocrine regulation exhibit age-related changes *(Finch, 1973; Brody, 1976; Simpkins et al., 1977; Riegle and Meites, 1976; Robinson, 1975; McGeer and McGeer, 1976)*.

The hypothesis of neuroendocrine regulatory alteration during aging has been a focus of increasing interest for several years *(e.g. Finch, 1973; Shock, 1974)*, but it has been developed most fully in an excellent theoretical review by Finch *(1976)*. According to this view, alterations in the set points of certain neuroendocrine control mechanisms during aging lead to initial endocrine imbalances which, in turn, lead to other physiological and metabolic imbalances and further alterations in control mechanisms, until the delicate balances of many physiological systems are disrupted by a "cascade" of metabolic disturbance which is perceived as senescence (e.g., atherosclerosis, diabetes, muscular wasting, immunological imbalances, etc). Finch *(1976)* has also suggested that neuroendocrine changes seen during aging may be gradual and progressive continuations of set point alteration mechanisms thought to be present at many stages of development, puberty and perhaps, maturity. These areas of research along with those dealing with immunological mechanisms appear to hold great promise for our understanding of "systemic" or "extrinsic" factors (i.e., factors not inherent to each cell) in the aging process.

In the present paper, however, I wish to consider data and concepts related to aspects of neuroendocrine alteration during aging which are not generally dealt with. Additionally, I wish to propose a new hypothesis; one which is conceptually less complex than that of neuroendocrine regulatory alteration, but one which may provide an important addition to our concepts on neural-endocrine interactions during aging; or, at least, will provide an hypothesis which can serve as a basis for empirical tests of whether such an addition to our concepts would be useful.

The experiments to be summarized here are relevant to neuroendocrine aging phenomena in two areas.

(1) Despite substantial evidence of decreased neurotransmitter functions, it is not clear how, or even if, these are translated into synaptic physiological deficits. That is, the decreased receptors alluded to above, or postsynaptic supersensitivity with age (which has been reported to occur in several peripheral somatic and sympathetic systems; e.g., *Eisdorfer, 1972; Vysocil and Gutmann, 1972; Korczyn, Laor and Nemet, 1976)* could conceivably partially compensate for neurotransmitter changes, and could thereby prevent major functional deficits. Thus, it seems important to study synaptic functioning during aging with neurophysiological, as well as with chemical and anatomical, methods. Below, I will summarize results of the first microelectrode studies of specific synaptic functions in the brains of aging animals, and consider how these findings may relate to present concepts.

(2) Another test of the neuroendocrine hypothesis which seems to me at least to be important, is the determination of whether or not age changes in any peripheral endocrine systems are *quantitatively* correlated with age-related brain changes. That is, if endocrine alterations are related to initial changes in brain control mechanisms, then there seemingly should be a somewhat proportional relationship between degree of brain aging, assessed by some quantitative index, and levels of activity in at least some peripheral endocrine systems. However, alterations in bioactivity without changes in quantity of some hormones, as described by

Adelman *(this volume)* might also play a role in deregulatory phenomena.

The possibility of a quantitative brain-endocrine correlation is also relevant to the hypothesis which I wish to propose here. There is another issue which has been raised by several participants in this Symposium: What is the initial cause of the neural changes leading to altered control mechanisms and, in turn, to endocrine and physiological imbalance? Although considerable evidence has been discussed at this Symposium to strongly suggest that dopaminergic systems are particularly susceptible to age effects, it has also been noted that other systems for which the neurotransmitters are unknown, or are non-aminergic, also show widespread deterioration *(e.g., Brody, 1973; Terry and Wisniewski, 1972; Scheibel and Scheibel, 1975; Vaughan, 1977; Geinisman, Bondareff and Dodge, 1977)*. Thus, the issue of the first cause of neural degeneration is important to the hypothesis of neuroendocrine regulatory alteration.

The hypothesis to deal with this issue which I wish to propose is that of the endocrine destruction of nerve cells during aging. I have suggested this elsewhere *(Landfield, unpublished; also see Landfield and Lynch, 1977b)*, but I would like to eleborate somewhat on this view in this paper. In particular, I wish to propose that the adrenocortical steroids (primarily glucocorticoids) are a major factor in effecting, or at least accelerating, the aging of the brain. However, there also seems to be a good possibility that all steroids, and perhaps thyroid hormones as well, induce erosion of their specific brain target cells.

An adrenocortical hypothesis of aging, of course, originated long before this paper. The studies of Wexler, Robertson and associates on the Pacific salmon, and with breeder rats, and the studies of Selye and colleagues on stress, as well as earlier clinical observations, led to the suggestion that elevated glucocorticoids were responsible for may physiological concomitants of aging *(reviewed in Wexler, 1976; Selye and Tuchweber, 1976)*. The contribution to these views I wish to make is to suggest: (a) that adrenocorticoids affect aging of *brain* cells; and (b) that elevated levels of corticoids may accelerate this process, but are not necessary to the effect; that is, that normal levels of corticoids can induce brain aging, but that normal levels will require a longer time to attain degrees of aging seen with elevated levels. This view predicts that a quantitative relationship should obtain between plasma levels of glucocorticoids and measures of brain aging, particularly in major target cells of the corticoids. The bases for these views is being reviewed extensively elsewhere *(Landfield, submitted)*.

The endocrine destruction of brain cells hypothesis is schematically illustrated in Figure 1 (solid lines), along with a simplified version of the neuroendocrine alteration hypothesis (dotted lines). Endocrine-induced aging of brain cells, if it exists, could be a result of initial brain deterioration arising from other

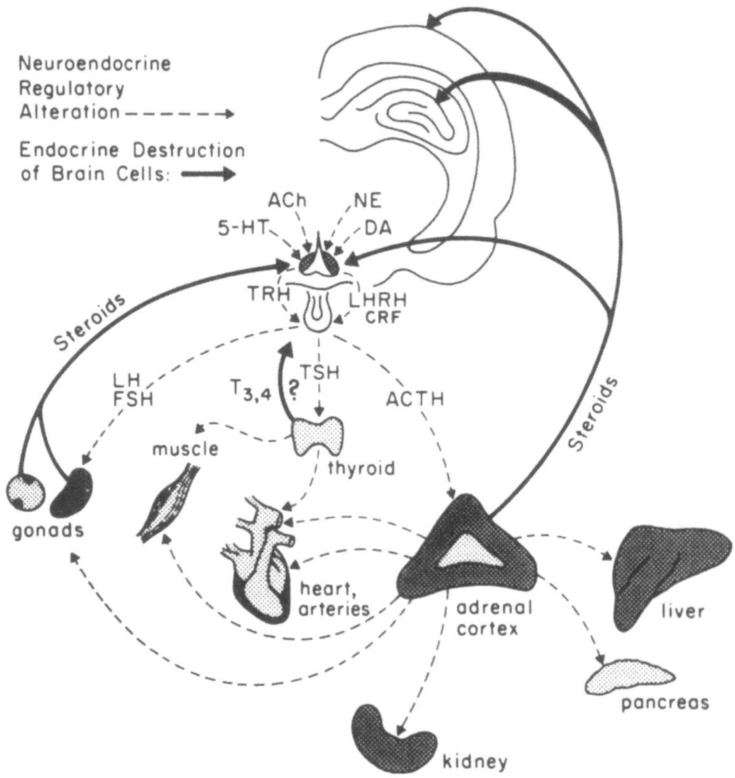

Figure 1. Schematic representation of two hypotheses of neural-endocrine interactions during aging; the neuroendocrine regulatory alteration hypothesis, in which changes in brain-endocrine control mechanisms are proposed to lead to peripheral physiological deterioration; (e.g., the neuroendocrine cascade hypothesis of Finch, 1976); and the hypothesis of endocrine acceleration of the rate of brain aging. Both mechanisms could operate together in a positive feedback loop.

sources (e.g., genetic, cardiovascular) leading to elevated endocrine activity and thus to additional loss of brain cells or synaptic function; alternatively, *normal* endocrine levels could be an early cause of brain cell deterioration, leading to elevation of adrenocroticoids, additional brain cell loss, additional deregulation, etc.; in other words, a "runaway positive feedback loop" between neuroendocrine regulatory alteration and endocrine-induced neural destruction could be formed.

In the following pages I will also describe initial correlative experiments which are consistent with both neuroendocrine regulatory alteration and endocrine brain destruction hypothesis of aging; at this time, clear evidence of adrenocortical age-like effects on brain cells is lacking, but I am presently conducting experiments which I hope will conclusively test this possibility.

In our neurophysiological studies, we have been using the hippocampus as a model system for the study both of aging and of memory. There is substantial, though controversial, evidence that the hippocampus of rodents is involved in "recent" memory processes *(reviewed in Landfield, 1976; 1978;* of course the evidence in humans is far more conclusive: e.g., *Milner, 1970)*. The present and ongoing work in these aging Fischer rats, which have been shown to exhibit substantial memory deficits *(Gold and McGaugh, 1975)*, greatly strengthens the view, I believe, of an involvement of rodent hippocampus in memory mechanisms. However, these points are developed more extensively elsewhere *(Landfield, 1978)*, and the present paper concentrates more on the implications of these studies for neuroendocrine mechanisms.

Our neurophysiological studies have so far only been conducted in hippocampus, both *in vivo*, in the acute, anesthetized (urethane) animal *(Landfield, McGaugh and Lynch, 1978)*, and in the *in vitro* hippocampal slice preparation *(Landfield and Lynch, 1977a)*. The neurophysiological studies utilized extracellular microelectrode recordings of synaptic responses which have been carefully analyzed by Andersen, Lomo and their associates† *(reviewed in Andersen, 1975)* and by others *(Fujita and Sakata, 1962; Deadwyler, West, Cotman and Lynch, 1975)*. These responses allow for extracellular analysis, not only of unit spikes but also, apparently, of summated EPSPs and IPSPs, because of the extremely dense hippocampal packing of cells. In Figure 2, the physiological preparations are schematically shown, along with the population EPSP (slow positive potential), population spike (fast negative spike), and population IPSP (slow positive potential following spike) of the typical postsynaptic response elicited by a single stimulation pulse administered to the Schaffer-commissural pathway. Considerable evidence supports the view that these are postsynaptic summations of unit EPSPs, spikes and IPSPs, induced by electrical stimulation of an input pathway, which exhibit this particular configuration of polarities due to a source-sink current relationship between cell somata and terminal fields on the dendrites *(Andersen, 1975)*. Many laboratories have demonstrated that the slice exhibits physiological responses generally similar to those seen *in vivo*. Our slice studies were carried out in a modified *in vitro* chamber, developed primarily in the laboratory of Gary Lynch, that allows for stability of the responses over many hours. The concomitant use of *in vitro* and *in vivo* preparations allows us to guard against the possibility that aged tissue might be more affected, on the one hand, by anesthesia, or, on the other, by slicing for *in vitro* studies.

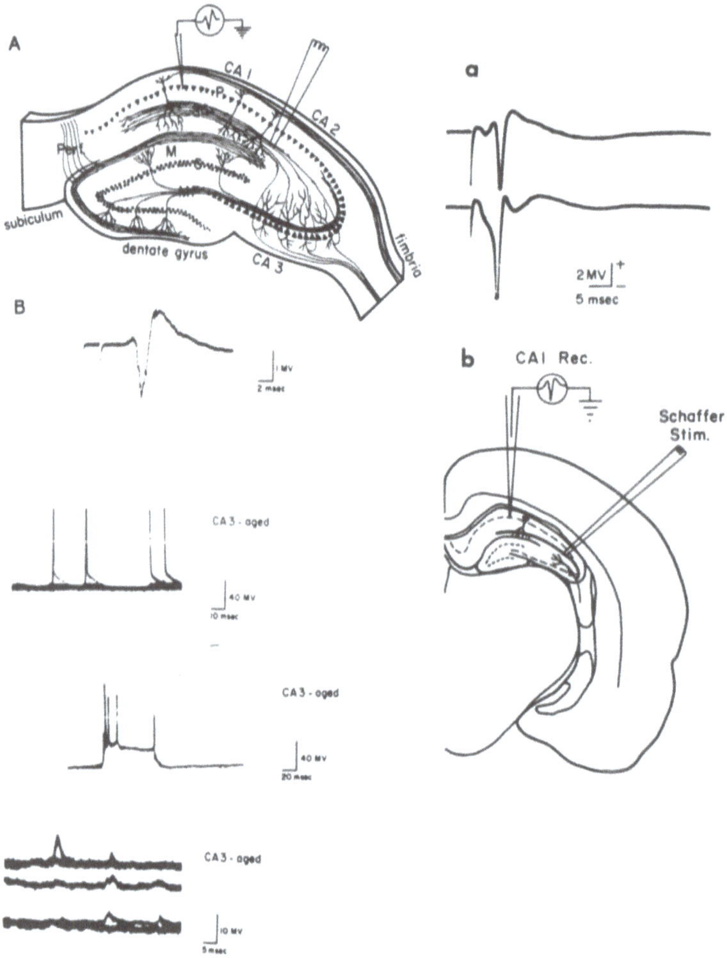

Figure 2. Diagram of the two neurophysiological preparations employed in the studies described. Top left: A, the hippocampal slice preparation, showing positions of the stimulating electrode in the Schaffer-commissural system in CA2-3 and the recording micropipette in the pyramidal cell body layer of CA1. B, the typical postsynaptic response recorded to a stimulation pulse given to the Schaffer collaterals. Right: a, the same response recorded from the intact animal. Upper trace from pyramidal somata; lower trace from the terminal field of the Schaffer collaterals on the apical dendrites of CA1 pyramids. Note polarity reversal of the EPSP component (please see text). b, schematic of the electrode positions in the intact preparation. The three traces on the lower left show intracellular records

To briefly summarize, we found no differences between young and aged hippocampal synapses in their evoked postsynaptic responses, either in terms of amplitude, latency or pattern, when stimulation frequencies were low (Figure 3). Further, no obvious differences were observed in spontaneous levels of spike activity.

The hippocampus exhibits forms of synaptic plasticity (e.g., frequency potentiation, paired-pulse facilitation, posttetanic potentiation; depression during electrical stimulation) which are similar to those seen in many other synaptic systems. However, some aspects of the responses in hippocampus are somewhat unique in that the frequency and posttetanic potentiation are extremely robust, with remarkably low thresholds and, moreover, under some conditions, the posttetanic potentiation (PTP) appears to be almost nondecremental, lasting in various reports for from hours and days to weeks *(Bliss and Lomo, 1973; Douglas and Goddard, 1975)*. In most other mammalian systems PTP is gone within minutes, or at most an hour. Thus, hippocampal PTP has been termed long-term potentiation (LTP) and has generated much interest in terms of possible relations to memory processes.

In our studies, in both preparations, we found that aged synaptic responses were only clearly deficient when challenged with higher frequency stimulation; that is, they were deficient in frequency and long-term potentiation and exhibited a greater tendency to exhaustion or depression (Figure 4). Such deficiencies were only seen during orthodromic but not during antidromic, activation. We have proposed that these deficits in hippocampal synaptic plasticity are relevant to age-related deficits in memory in these rats *(Landfield, 1978; Landfield, McGaugh and Lynch, 1976; Landfield and Lynch, 1977)*. Moreover, in ultrastructural studies my associates and I have very recently obtained evidence that this synaptic deficit is related to a quantitative age-related reduction in populations of synaptic vesicles in hippocampal synapses.

However with regard to possible implications for neuroendocrine regulation during aging, the following points may perhaps be noted. First, these functional data are highly consistent with chemical studies showing neurotransmitter changes with age and suggest that the brain might well be altering or losing adequate control over hypothalamic hypophysiotropic systems. Second, loss of synaptic control would appear greatest during higher frequency activation of neural systems. This finding seems highly consistent with a number of studies showing unimpaired baseline hormone

---

from a CA3 pyramidal cell in a hippocampal slice from an aged animal. Note that the top intracellular trace shows spontaneous action potentials of 100 mV amplitudes. Figures reproduced with the kind permission of the editors of J. Gerontology (A and B) and Brain Research (a and b). Intracellular recordings obtained in collaboration with Dr. Sam Deadwyler.

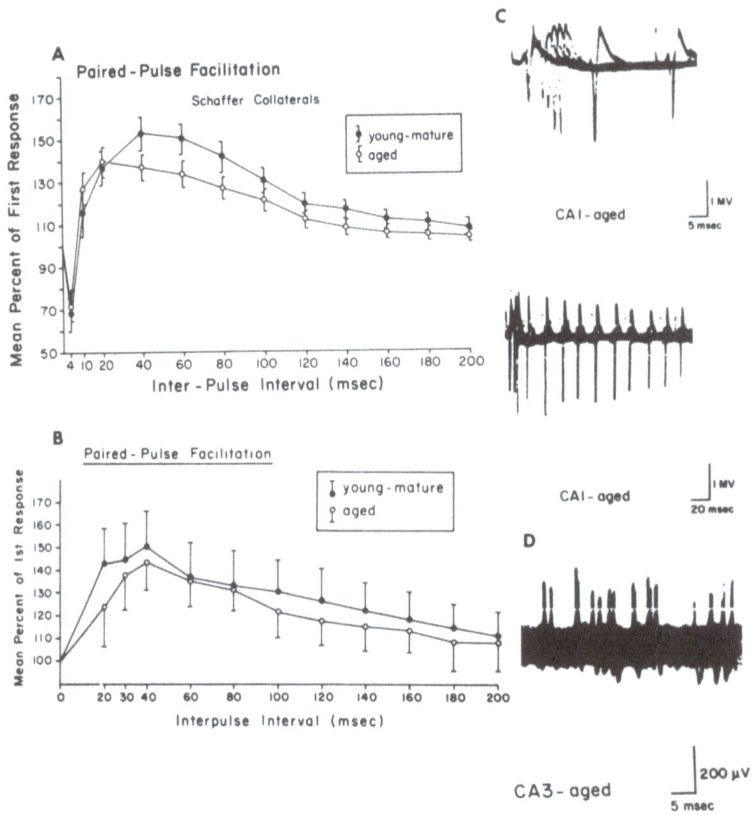

Figure 3. Paired--ulse facilitation curves showing no major differences between aged and young animals, either in vitro (A) or in vivo (B). Graphs are of the amplitudes of the postsynaptic response to the second pulse of a pair of stimulation pulses at the same intensity, plotted as percent of the response to first pulse, as a function of interpulse interval. C: second responses of a pair superimposed on a storage oscilloscope as the interpulse interval is varied. D: apparently normal extracellular unit spike activity recorded from an aged slice. Spikes are superimposed on a storage scope.

Figure 4. Effects of repetitive stimulation. Left column: impaired frequency potentiation (growth of a response during train of stimulation pulses) (15 Hz for in vitro study; 6 and 12 Hz for in vivo study) in aged hippocampus both in vitro (a) and in vivo (b). Right column:

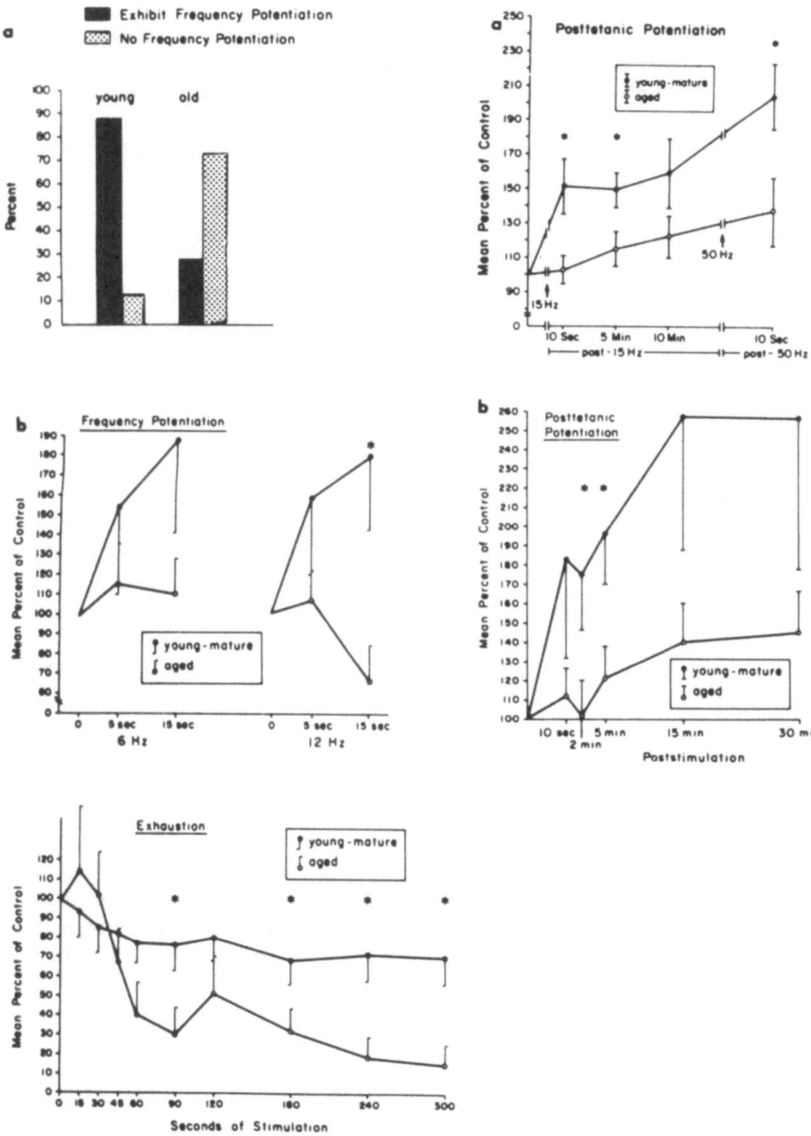

Figure 4. (cont'd) *impaired posttetanic, or long-term, potentiation (enhancement of responses to a single pulse following a train of stimulation pulses) (10 sec of 15 and 50 Hz, for in vitro; 5 sec of 100 Hz, in vivo), in aged hippocampus, both in vitro (a) and in vivo (b). Lower figure shows increased exhaustion or fatigue at aged hippocampal synapses in the intact preparation, during continuous 4 Hz stimulation for 5 min. Reproduced with permission of the editors of J. Gerontology (a,a) and Brain Research (b,b, and lower figure).*

levels, but sluggish endocrine responses to external stimuli, in aging animals *(reviewed in Adelman, 1976; Finch, 1976)*. In fact, the rates of most neuromuscular responses to challenges are reduced during aging *(Shock, 1961, 1974; Birren, 1965)*. Moreover, the inability of hippocampal synapses to respond normally to high frequency stimulation seems of interest in light of the correlation between high frequency activity in neuroendocrine cells and release of hormones *(Hayward and Jennings, 1973; Dyball and Dyer, 1971)*. Lastly, the hippocampus appears to play a role in the regulation of ACTH secretion, particularly in relation to stress. Under *some* conditions, this role has been reported to be inhibitory *(reviewed in Bohus, 1975)*. Therefore, it seems conceivable that deteriorating hippocampal synaptic functions, according to the neuroendocrine deregulation hypothesis, could be associated with alterations in adrenocortical activity.

This possibility, along with evidence that the hippocampus is a major target organ for corticosterone *(reviewed in McEwen, Gerlach and Nicco, 1975; Dekloet and McEwen, 1976)*, and the studies of Wexler and associates, noted earlier, led to the experiment described below *(Landfield, Waymire and Lynch, unpublished observations)*. Radioimmunoassays for plasma corticosterone and aldosterone, measures of adrenal weights, and quantitative analyses of the degree of hippocampal aging were carried out in 9 young (4 mo-old), 9 mid-aged (13 mo-old) and 9 aged (25 mo-old) inbred Fischer 244 rats, obtained from a germ-free colony maintained at Charles River Co. under contract to the National Institute of Aging and Dr. D. Gibson.

The animals were sacrificed in triplets (1 young, 1 mid-aged, 1 aged) over a period of 5-6 hours, due to extensive dissection of organs from these expensive animals for use in other studies. However, the pattern of corticoid variations, the paired controls for time of sacrifice, and the magnitude of variation, all suggested that individual rather than diurnal factors accounted for the largest amounts of variance.

Because there is considerable variance in the neurophysiological measurements, we used an anatomical rather than a neurophysiological index of hippocampal aging. This index was a quantitative analysis of the substantial increase in gold chloride staining of reactive astrocytes which occurs in hippocampus of aging Fischer rats *(Landfield, Rose, Sandles, Wohlstadter and Lynch, 1977)*. This increase in reactive astrocytes is shown in Figure 5.

It seems clear that this age-related astrocyte reactivity (which is most prominent in, but is not limited to, hippocampus) was not due simply to a change in the staining properties of aging astrocytes. We are conducting quantitative ultrastructural studies of this astrocytosis and, although these are not completed, it is clear that there is a substantial increase in the number of astroglial profiles in aged tissue. Moreover, the aged astrocytes contain considerable numbers of inclusions which are very different from the classical lipofuscin bodies observed in neurons and in peripheral tissues.

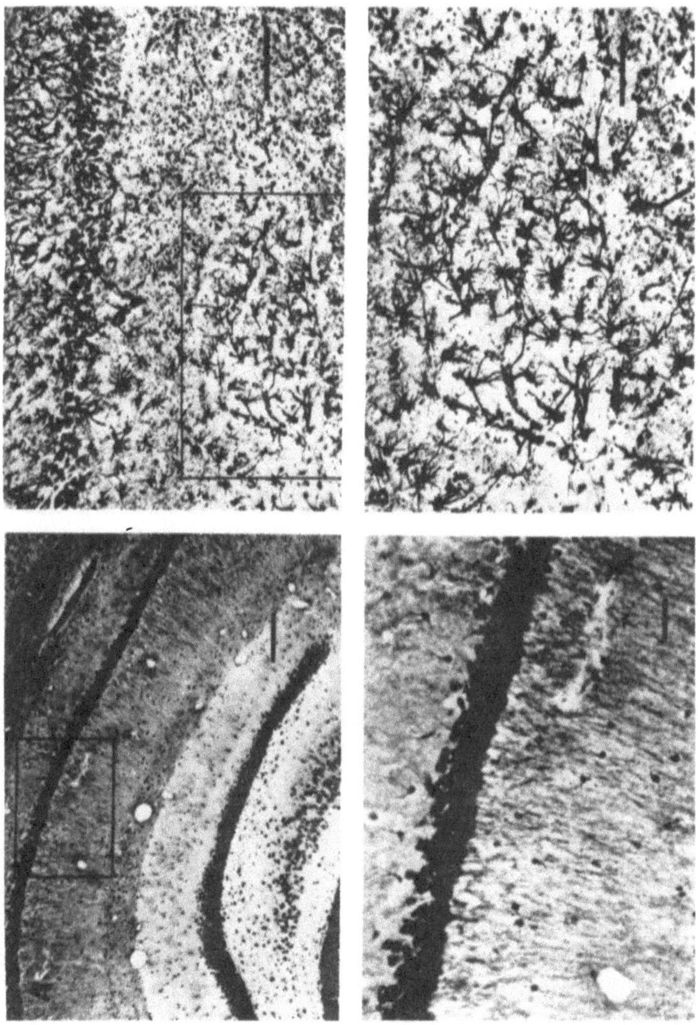

Figure 5. *Gold chloride stains for astrocytes in young (left) and aged (right) Fischer rats. The lower figures (B) are blow-ups of the outlined regions in the upper figures (A). Calibration: Left: A, 200μ; B, 50μ; Right: A, 100μ; B, 50μ. Note that the astrocytes are far more reactive in the aged animals, and that the neuronal somata of pyramidal cells also stain more weakly in the aged animals (cell layer running across figure A, left and right sides). Reproduced with permission of the publishers of J. Gerontology.*

In many ways, these astroglial inclusions, similar to those reported by Vaughan and Peters *(1974)* in cortical microglia of aging rats, resemble the phagocytosed synaptic contacts seen in experimental degeneration studies *(Mugnaini, Walberg and Brodal, 1967; Cook and Wisniewski, 1973)*. Additionally, we have recently obtained good evidence that the major factor involved in triggering the astroglial reactivity is axon terminal degeneration (Figure 6C).

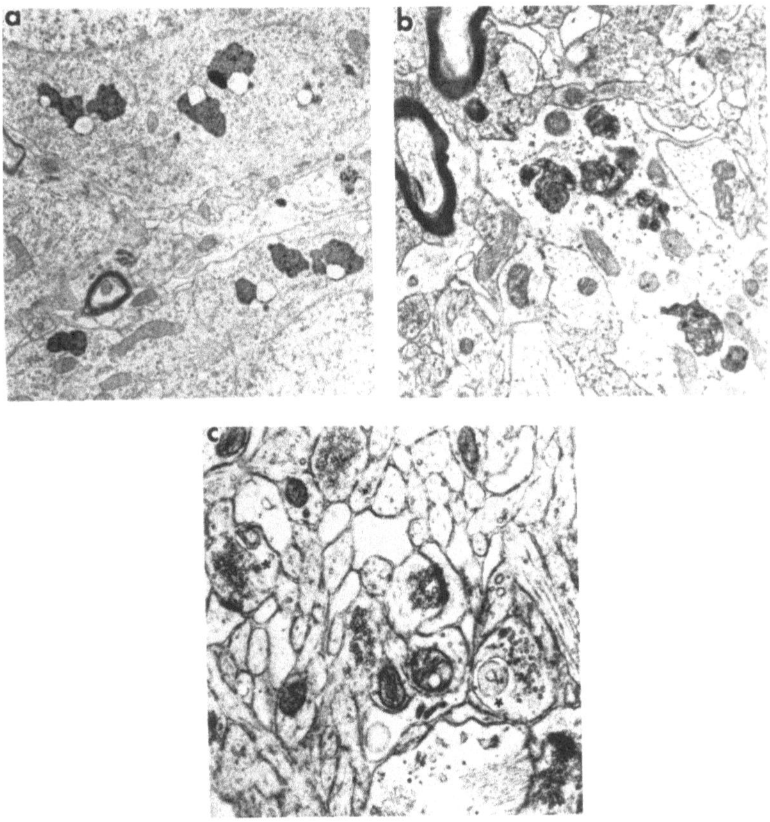

Figure 6. Electron micrographs from hippocampal formation of an aged Fischer rat. a: typical lipofuscin bodies in the perikarya of dentate gyrus granule neurons. b: membranous inclusion bodies in an astrocytic process from the neuropil of the Schaffer collateral synaptic terminal field. c: micrograph from the same region of another aged animal showing membranous vacuolar degeneration in axon terminals (stars), possibly surrounded by astrocytic processes. These degenerating elements could be related to mitochondria. Note large reactive astrocytic process in lower right, containing filament and glycogen granules.

Thus, these glial alterations may reflect degenerative processes similar to those observed in human senile plaques *(Terry, Wisniewski, 1972; Wisniewski, Ghetti and Terry, 1973)*.

In any case, the reactive astrocytes appear to be excellent correlates of age changes in hippocampus since a change can be observed in mid-age, and since astrocyte reactivity is a well-known general correlate of brain damage or of brain edema.

Analysis was performed by a camera lucida method on a Zeiss photomicroscope, using a drawing tube; the hippocampal region which was quantified in a "blind" study is shown in Figure 7. A grid reticule was superimposed over this area and total cells along with reactive cells were quantified in 9 mid-aged, 6 young and 6 aged animals. (Neurons of the pyramidal cell layer were not included). All counts were corrected for size of the delimited region by dividing these by the number of grid squares covering the area. Table 1 shows the results of this study in terms of age group means.

It can be seen in Table 1 that significant increases in plasma adrenocorticoids (tests for paired samples, to control for diurnal effects) and in reactive astrocytes have developed by 13 mo in these Fischer 344 animals (mean life span in various reports is from 20-29 mo). Total cells (including reactive astrocytes, nonreactive astrocytes, microglia, oligodendroglia and interneurons) tend to increase in the aged animals, but this is not significant due to large variance. However, a number of other investigators have reported increases in glial numbers with age, at least in some regions *(e.g., Brizzee, Sherwood and Timiras, 1968; Vaughan and Peters, 1974)*. Adrenal weight is significantly elevated in the aged animals, a pattern suggesting perhaps that some adrenal stim-

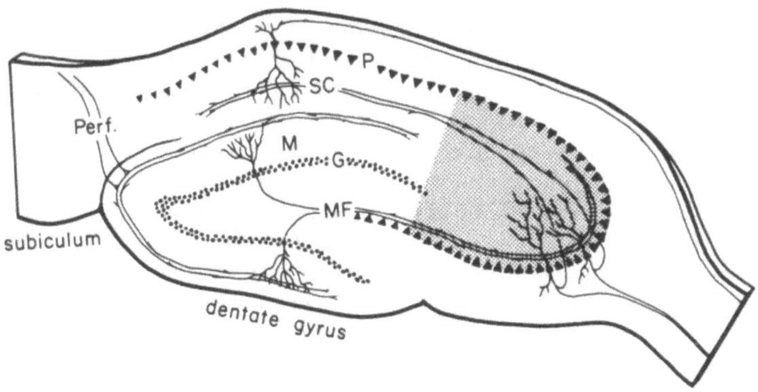

Figure 7. Schematic representation of a cornual slice of hippocampus. The region quantified for reactive astrocytes is shown in stippling. P: pyramidal cell somata: SC: Schaffer collaterals; M: molecular layer of the dentate gyrus; G: granule cell somata layer; MF: mossy fibers; Perf: perforant path fibers from entorhinal cortex.

TABLE 1. ADRENAL FUNCTION AND MEAN DENSITY OF REACTIVE AND TOTAL CELLS (PER GRID SQUARE) WITHIN THE QUANTIFIED REGION FOR THE 3 AGE GROUPS.

	Reactive Cells	Total Cells	Corticosterone (ng/ml)	Adrenal weight (mg/pair)
4 mo.				
mean ± 1 S.E.M.	1.21	20.7	42.0	37.7
(n=6)	±.25	±2.5	±10.6	±2.8
13 mo.				
mean ± 1 S.E.M.	3.19	20.1	71.2	39.9
(n=9)	±.45	±1.3	±14.0	±7.2
25 mo.				
mean ± 1 S.E.M.	7.02	23.7	81.6	52.5
(n=6)	±1. 1	±2.2	±29.0	±8.3

ulating factor develops by 13 mo and leads to glandular hypertrophy by 25 mo.

Lewis and Wexler *(1974)* reported elevated corticosterone in 19 mo-old virgin male rats, along with hypertrophied adrenals which stained heavily for lipids. Our 25 mo-old animals exhibit enlarged adrenals, which also stain heavily with Sudan Black for neutral lipids (Figure 8), but they do not exhibit elevated steroids. That is, median values are almost the same as in the young group, and the mean elevation is accounted for by two aged animals with values well above any seen in young or mid-aged animals. This latter observation supports the interpretation of an initial elevation of adrenal function in mid-age followed by a subsequent "burning out" or failure of adrenal function in most animals. Riegle and Hess *1972)* reported an elevated threshold for ACTH suppression by steriods with age, both of which might fit some aspects of our findings. Das and Magilton *(1971)* also reported elevated adrenal weight in conjunction with body weight in aging dogs.

Thus, the apparent biphasic pattern of initial elevation and subsequent reduction of adrenocorticoids which we observed could account for some of the conflicting data in the literature. Of course, strain, species and relative physiological age differences also no doubt play a role.

The changes in mean values for the three age groups seem to provide interesting data relevant to the issues at hand. However,

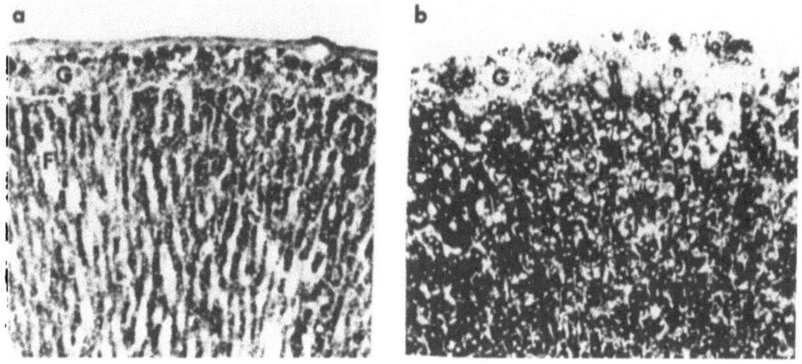

Figure 8. Sudan Black B stain for lipids of the adrenal cortices from young (a) and aged (b) rats. G: zona glomerulosa; F: zona fasiculata. There is greater lipid staining in the aged zona fasiculata.

correlations among individual subjects provide a more powerful analytical tool, and were the main purpose of the study. In aged animals, the pattern of adrenal changes suggested that steroid levels found might not be those which had obtained throughout most of the animal's lifespan. Conversely, the young animals do not exhibit sufficient reactive gliosis to attempt an individual subject correlation. Therefore, the correlation study primarily focused on the mid-aged group. The results of the correlation between plasma corticosterone and summed gliosis (a measure giving approximately equal weight to reactive cells and total cells) are shown in Figure 9. This correlation was r = +.76. The correlation of corticosterone with reactive cells alone was also significant (r = +.65). The correlation of reactive cells alone in the 6 aged animals in which astrocytes were quantified with *adrenal weight* was significant (r = +.71), but was not quite significant with the measure of summed gliosis (r =+.65). Aldosterone exhibited positive though nonsignificant correlations with measures of brain gliosis.

The correlation of adrenal weight with corticosterone in aged animals suggests that, in 25 mo-old animals, adrenal weight is a better measure of the stimulation to which the gland had been subjected to than are plasma steroids, and, moreover, supports the plasma steroid-gliosis correlation observed in the mid-aged group (*Addendum*).

An interesting aspect of these studies was a significant correlation observed between plasma corticosterone and plasma pH in these 13 mo and 25 mo-old animals (Figure 10).

Although few studies have reported pH alterations during aging, such studies have usually been conducted *in situ*. In the present case, however, plasma was analyzed for pH after CO_2 had evaporated, and after the hemoglobin had been separated by centrifugation from

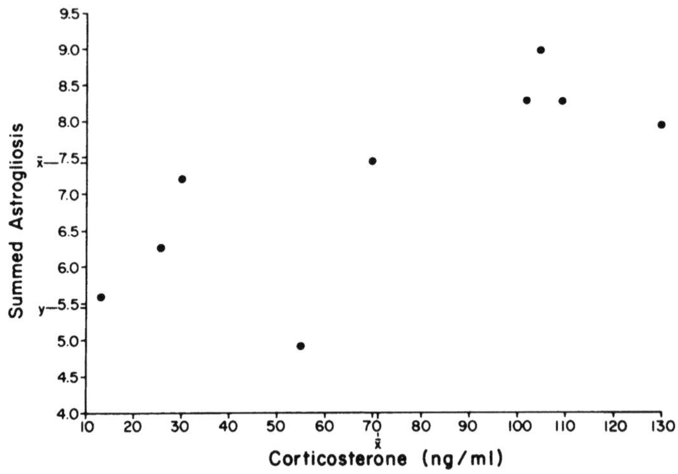

Figure 9. Correlation in mid-aged (13 mo-old) rats between a quantitative measure of astrogliosis in the hippocampus and plasma levels of corticosterone in the same animals.

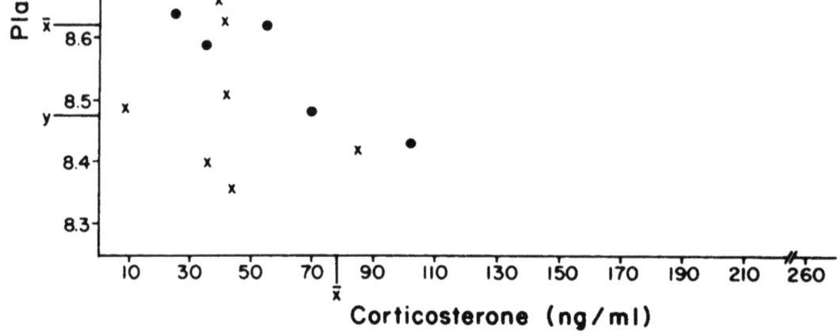

Figure 10. Correlation in aged and mid-aged rats between plasma corticosterone and plasma pH. The latter measure was obtained after centrifugation and CO_2 evaporation, and therefore buffer systems were reduced. Four mo-old animals exhibited a similar correlation.

plasma. This loss of important buffering factors is the reason for the generally elevated pH observed. However, the correlation with corticosterone levels (but not with aldosterone) is more difficult to account for.

Corticosterone is known to have a direct mobilizing and depleting effect on cellular potassium which is independent of the kidney effects of adrenocorticoids. In extreme cases, this can lead to extracellular hypokalemic alkalosis (and intracellular acidosis). It seems conceivable that, when uncovered by loss of buffering capacities, a relationship between individual corticosterone levels and plasma ionic composition might be observed. The possibility, from another viewpoint, that intracellular accumulation of K^+ could affect DNA function and cellular hydration during aging has also recently been reported *(Zs-Nagy, Zs-Nagy, Pieri, Giulli and DelMoro, 1977)*.

Intracellular microelectrode analyses of membrane resting and action potentials in brain neurons seems to be one means of determining whether intracellular ionic composition is disturbed with age. However, many cells would have to be analyzed statistically to determine if age-related differences exist. Nevertheless, I will briefly report on some extremely preliminary data collected from only a few cells from aged and young hippocampal slices. In this preparation, the extracellular ionic composition can be held fairly constant (leaving aside possible ionic sequestering in the extremely narrow hippocampal extracellular spaces) and differences should be largely attributable to intracellular ionic composition. These preliminary studies were conducted in collaboration with Dr. Sam Deadwyler.

In several aged slices, action potentials approximating 100 mv were recorded (Figure 2, lower left traces). Spikes of this size were rarely observed in slices from young animals. Moreover, in *in vivo* studies in the cat, Kandel, Spencer and Brinley *(1961)* rarely recorded action potentials of this size from hippocampal pyramidal cells. We have not observed enough cases to draw any firm conclusions. Nevertheless, it seems of interest that, if true, these findings could be accounted for by intracellular accumulation of K^+ (increased membrane resting potential) or by depletion of intracellular Na^+ (increased action potential), ionic alterations which might occur if adrenal functions were failing in the aged animal. However, possible consequences of ionic alterations, induced by steroid changes, on brain aging processes are highly speculative at this point.

In summarizing, I believe that the neurophysiological data, along with the brain-endocrine correlations, are consistent with previously untested predictions arising from the neuroendocrine regulatory alteration hypothesis, and strengthen this concept. Additionally, however, the correlative data are consistent with an hypothesis of endocrine acceleration of brain aging through hormonal actions on brain target cells, by as yet unspecified mechanisms. The latter is a conceptually simple and testable hypothesis, which I have begun to examine, and which, hopefully, may facilitate the analysis of neural-endocrine interactions in the aging process.

Addendum: Note added in proof. We have just completed a study of the effects of prolonged administration of adrenal hormones on hippocampal astroglial reactivity. These studies clearly confirm the adrenal weight-astrogliosis correlation in untreated, aged rats ($r = +.76$, $n = 10$). Moreover, rats treated with hormones exhibited patterns of brain pathology fully consistent with the hypothesis of glucocorticoid acceleration of brain aging (Figure 1) as described by Landfield, Lindsey, Lynch *(1978)*.

ACKNOWLEDGEMENTS

Much of the research discussed here was conducted in association with a major collaborator, Dr. Gary Lynch, and with Drs. James McGaugh and Jack Waymire. Additionally, several student associates, James Lindsey, Kevin Lee and Greg Rose, have made substantial contributions. Aspects of this research were supported by NIH grants Ag00341 to the author, Ag00538 to G. Lynch and J. Waymire, Ag00439 to J. McGaugh and NSF grant BMS 7202237 to G. Lynch.

REFERENCES

Adelman, R.C. (1976). Age-dependent hormonal regulation of mammalian gene expression. In: *Hypothalamus Pituitary and Aging*, pp. 668-675, eds. A.F. Everitt and J.A. Burgess, Charles C. Thomas, Springfield, Ill.

Andersen, P. (1975). Organization of hippocampal neurons and their interconnections. In: *The Hippocampus*, pp. 155-175, eds. R. L. Isaacson and K.H. Pribram, Plenum Press, New York.

Birren, J.E. (1965). Age changes in speed of behavior: Its central nature and physiological correlates. In: *Behavior, Aging, and the Nervous System*, eds. A.T. Welford and J.E. Birren, Charles C. Thomas, Springfield, Ill.

Blackwell, R.E. and Guillemin, R. (1973). Hypothalamic control of adenohypophysial secretions. *Ann. Rev. Physiol., 35*, 357-390.

Bliss, T.V.P. and Lømo, T. (1973). Long lasting potentiation of synaptic transmission in dentate area of the anesthetized rabbit following stimulation of the perforant path. *J. Physiol. 232*, 331-356.

Bohus, B. (1975). The hippocampus and the pituitary-adrenal system hormones. In: *The Hippocampus, vol. 1*, pp. 323-353, eds. R.L. Isaacson and K.H. Pribram, Plenum Press, New York.

Brizzee, K.R., Sherwood, N. and Timiras, P.S. (1968). A comparison of cell populations at various depth levels in cerebral cortex of young adult and aged Long-Evans rats. *J. Geront.* 23, 289-297.

Brody, H. (1973). Aging of the vertebrate brain. In: *Development and Aging in the Central Nervous System*, pp. 121-133, eds. M. Rockstein and M. Sussman, Academic Press, New York.

Brody, H. (1976). An examination of cerebral cortex and brainstem aging. In: *Neurobiology of Aging, 3*, pp. 171-181, eds. R.D. Terry and S. Gershon, Raven Press, New York.

Clemens, J.A. and Bennett, D.R. (1977). Do aging changes in the preoptic area contribute to loss of cyclic endocrine function? *J. Geront. 32*, 19-24.

Cook, R.D. and Wisniewski, H.M. (1973). The role of oligodendroglia and astroglia in Wallerian degeneration of the optic nerve. *Brain Res. 61*, 191-206.

Das, L.N. and Magilton, J.H. (1971). Age changes in the relationship among endocrine glands of the beagle. *Exp. Geront. 6*, 313.

Deadwyler, S.A., West, J.R., Cotman, C.W. and Lynch, G.S. (1975). A neurophysiological analysis of the commissural projections to the dentate gyrus of the rat. *J. Neurophysiol. 38*, 167-184.

DeKloet, R. and McEwen, B.S. (1976). Glucocorticoid interactions with brain and pituitary. In: *Molecular and Functional Neurobiolgoy*, pp. 257-295, ed. W.H. Gispen, Elsevier, Amsterdam.

Douglas, R.M. and Goddard, G.V. (1975). Long-term potentiation of the perforant path-granule cell synapse in the rat hippocampus. *Brain Res. 86*, 205-215.

Dyball, R.E.J. and Dyer, R.G. (1971). Plasma oxytocin concentration and paraventricular neurone activity in rats with diencephalic islands and intact brains. *J. Physiol., Lond. 216*, 227-235.

Eisdorfer, C. (1972). Autonomic changes in aging. In: *Aging and the Brain*, ed. C.M. Gaitz, Plenum Press, New York.

Finch, C.E. (1973). Catecholamine metabolism in the brains of aging male mice. *Brain Res. 52*, 261-276.

Finch, C.E. (1976). The regulation of physiological changes during mammalian aging. *Q. Rev. Biol. 51*, 49-83.

Fujita, Y. and Sakata, H. (1962). Electrophysiological properties of CA1 and CA2 apical dendrites of rabbit hippocampus. *J. Neurophysiol. 25*, 209-222.

Geinisman, Y., Bondareff, W. and Dodge, J. (1977). Partial deafferentation of neurons in the dentate gyrus of the senescent rat. *Brain Res. 134*, 541-554.

Gold, P.E. and McGaugh, J.L. (1975). Changes in learning and memory during aging. In: *Neurobiology of Aging*, eds. J.M. Ordy and K.R. Brizzee, Plenum Press, New York.

Hayward, J.N. and Jennings, D.P. (1973). Activity of magnocellular neuroendocrine cells in the hypothalamus of unanesthetized monkeys. I. Functional cell types and their anatomical distribution in the supraoptic nucleus and the internuclear zone. *J. Physiol. Lond. 232*, 515-543.

Hess, G.D. and Riegle, G.D. (1970). Adrenocortical responsiveness to stress and ACTH in aging rats. *J. Geront. 25*, 354-358.

Kandel, E.R., Spencer, W.A. and Brinley, F.J. (1961). Electrophysiology of hippocampal neurons. I. Sequential invasion and synaptic organization. *J. Neurophysiol. 24,* 225-242.

Korczyn, A.D., Laor, N. and Nemet, P. (1976). Sympathetic pupillary tone in old age. *Arch. Ophthal. 94,* 1905.

Landfield, P.W. (1976). Synchronous EEG rhythms: Their nature and their possible functions in memory, information transmission and behavior. In: *Molecular and Functional Neurobiology,* pp. 390-424, ed. W.H. Gispen, Elsevier, Amsterdam.

Landfield, P.W. (1978). Composite memory, the hippocampus and aging. *Neuroscience and Biobehavioral Reviews,* in press.

Landfield, P.W. and Lynch, G. (1977a). Impaired monosynaptic potentiation in *in vivo* hippocampal slices from aged, memory-dedicient rats. *J. Geront. 32,* 523-533.

Landfield, P.W. and Lynch, G. (1977b). Brain aging and plasma steroids: Quantitative correlations. *Soc. Neurosci. Absts.* November, 1977.

Landfield, P.W., Lindsey, J.D., and Lynch, G. (1978). Apparent acceleration of brain aging pathology by prolonged administration of glucocorticoids. *Soc. Neurosci. Absts. Vol 4,* 350.

Landfield, P.W., McGaugh, J.L. and Lynch, G. (1978). Impaired synaptic potentiation processes in the hippocampus of aged, memory-deficient rats. *Brain Res.*, 150, 85-101.

Landfield, P.W., Rose, G., Sandles, L., Wohlstadter, T. and Lynch, G. (1977). Patterns of astroglial hypertrophy and neuronal degeneration in the hippocampus of aged, memory-deficient rats. *J. Geront. 32,* 3-12.

Lewis, B.K. and Wexler, B.C. (1974). Serum insulin changes in male rats associated with age and reproductive activity. *J. Geront. 29,* 139-144.

McGeer, E. and McGeer, P.L. (1976). Neurotransmitter metabolism in the aging brain. In: *Neurobiology of Aging,* pp. 389-403, eds. R.D. Terry and S. Gershon, Raven Press, New York.

McEwen, B.S., Gerlach, J.L. and Micco, D.J. (1975). Putative glucocorticoid receptors in hippocampus and other regions of the brain. In: *The Hippocampus, vol. 1,* pp. 285-322, eds. R.L. Isaacson and K.H. Pribran, Plenum Press, New York.

McIlwain, H. (1972). Electrical stimulation of specified subsystems of the mammalian brain, As isolated tissue preparations. In: *Experimental Models of Epilepsy,* pp. 269-289, eds. D.P. Purpura and R. Walter, Raven Press, New York.

Milner, B. (1970). Memory and the medial temporal regions of the brain. In: *Biology of Memory,* eds. K.H. Pribram and D.E. Broadbent, Academic Press, New York.

Mugnaini, E., Walberg, R. and Brodal, A. (1976). A mode of termination of primary vestibular fibres in the lateral vestibular nucleus. An experimental electron microscopical study in the cat. *Expl. Brain Res. 4,* 187-211.

Reichlin, S. (1974). Neuroendocrinology. In: *Textbook of Endocrinolocy, 5th Edition,* pp. 774-831, ed. R.H. Williams, Saunders, Philadelphis.

Riegle, G.D. and Hess, G.D. (1972). Chronic and acute dexamethasone suppression of stress activation of the adrenal cortex in young and aged rats. *Neuroendocrinology 9*, 175-187.

Riegle, G.D. and Meites, J. (1976). Effects of aging on LH and prolactin after LHRH, L-dopa, methyl-dopa, and stress in male rats. *Proc. Soc. exp. Biol. Med. 151*, 507-511.

Robinson, D.S. (1975). Changes in monoamine oxidase and monoamines with human development and aging. *Fedn. Proc. 34*, 103-107.

Schally, A.V., Arimura, A. and Kastin, A.J. (1973). Hypothalamic regulatory hormones. *Science 179*, 341-350.

Scheibel, M.E. and Scheibel, A.B. (1975). Structural changes in the aging brain. In: *Aging I*, pp. 11-37, eds. H. Brody, D. Harmon and J.M. Ordy, Raven Press, New York.

Selye, H. And Tuchweber, B. (1976). Stress in relation to aging and disease. In: *Hypothalamus, Pituitary and Aging*, pp. 553-569, eds. A.F. Everitt and J.A. Burgess, Charles C. Thomas, Springfield, Illinois.

Shock, N.W. (1974). Physiological aspects of aging in man. *Ann. Rev. Physiol. 23*, 97-122.

Shock, N.W. (1974). Physiological theroies of aging, In: *Theoretical Aspects of Aging*, ed, N. Rockstein, Academic Press, New York.

Simpkins, J.W., Mueller, G.P., Huang, H.H. and Meites, J. (1977). Evidence for depressed catecholamine and enhanced serotonin metabolism in aging male rats: Possible relation to gonadotropin secretion. *Endocrinology 100*, 1672-1678.

Terry, R.D. and Wisniewski, N.M. (1972). Ultrastructure of senile dementia and of experimental analogs. In: *Aging and the Brain*, ed. C.M. Gaitz, Plenum Press, New York.

Vaughan, D.W.(1977). Age-related deterioration of pyramidal cell basal dendrites in rat auditory cortex. *J. Comp. Neurol. 171*, 501-515.

Vaughan, D.W. and Peters, A. (1974). Neuroglial cells in the cerebral cortex of rats from young adulthood to old age: An electron microscope study study. *J. Neurocytol. 3*, 405-239.

Vyskocil, F. and Gutmann, E. (1972). Spontaneous transmitter release from nerve endings and contractile properties in the soleus and diaphragm muscles of senile rats. *Experientia 28*, 280-281.

Wexler, B.C. (1976). Comparative aspects of hyperadrenocorticism and aging. In: *Hypothalamus Pituitary and Aging*, pp. 333-361, eds. A.F. Everitt and J.A. Burgess, Charles C. Thomas, Springfield, Illinois.

Wisniewski, H.M., Ghetti, B. and Terry, R.D. (1973). Neuritic (senile) plaques and filamentous changes in aged rhesus monkeys. *J. Neuropath. exp. Neurol. 32*, 566-584.

Zs-Nagy, I., Zs-Nagy, V., Pieri, C., Giulli, C. and DelMoro, M. (1978). *In vivo* stimulation of nerve cells by phytohemagglutinin. *Gerontology 24*, 12-26.

CYCLIC NUCLEOTIDES IN NEUROENDOCRINE FUNCTION

YVONNE CLEMENT-CORMIER

Department of Pharmacology
The University of Texas Medical School
Houston, Texas 77025

Several antipsychotic agents produce extrapyramidal side effects which resemble Parkinson's disease *(Hollister, 1972; Hornykiewicz, 1973)*. It is generally agreed that these side effects may arise from the ability of these drugs to block the dopamine receptor of the caudate nucleus *(Carlsson and Lindqvist, 1963; Nyback and Sedvall, 1968)*. Biochemical and pharmacological studies have suggested an intimate association between the dopamine receptor and a dopamine-sensitive adenylate cyclase in the caudate since the properties of the enzyme mimicked in large part the actions of the dopamine receptor *(Kebabian, Petzold and Greengard, 1972; Clement-Cormier, Kebabian, Petzold and Greengard, 1974; Iverson, 1975; Andén and Stock, 1973)*.

Catecholamines, especially dopamine have been implicated in controlling certain aspects of neuroendocrine function *(MacLeod 1969; Birge, Jacobs, Hammer and Daughaday, 1970; Koch, Lu and Meites, 1970; Kamberi, Mical and Porter, 1971; Shaar and Clemens, 1974; Quijada, Illner, Krulich and McCann, 1973/74)*. The median eminence has been identified as one such area receiving dopaminergic innervation *(Kavanagh and Weisz, 1973)*. Since previous studies had supported the correlation between dopaminergic innervation of the extrapyramidal motor areas and the limbic system and the occurrence of dopamine sensitive adenylate cyclase in these areas, it was of interest to verify the presence of a dopamine-sensitive adenylate cyclase in the median eminence. It has been proposed that blocking the dopamine stimulation of adenylate cyclase by antipsychotic drugs in the caudate may relate in part to the ability of these agents to precipitate extrapyramidal side effects *(Martin, 1973)*. It has also been suggested that the endocrinological side effects of antipsychotic drugs, especially hyperprolactinemia, may result from a blockade of dopamine receptors. For these reasons it was of inter-

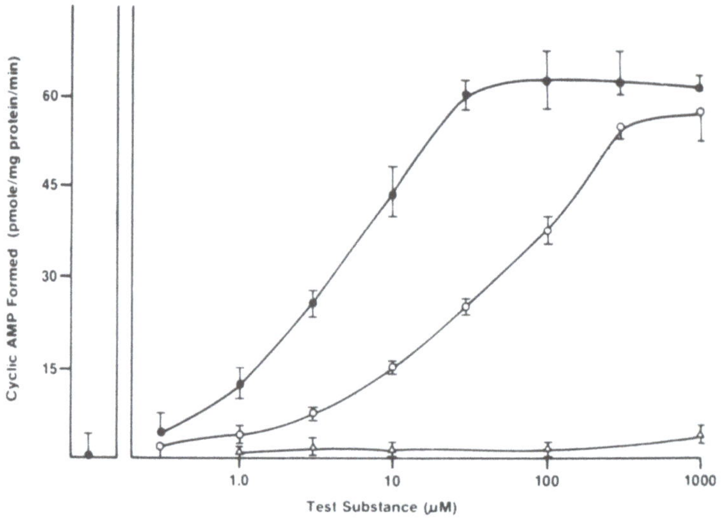

Figure 1. Effect of catecholamines on adenylate cyclase activity in a homogenate of rat median eminence. Standard conditions were used for the measurement of adenylate cyclase activity. In the absence of added catecholamine, 78.4 ± 0.9 pmoles (mg of protein)$^{-1}$ min^{-1} of cyclic AMP was formed. The increase in cyclic AMP above this basal level is platted as a function of catecholamine concentration. The data give the mean values and ranges for duplicate determinations on each of three replicate samples.

Key: (•) dopamine; (o) norepinephrine; (Δ) isoproterenol

est to investigate the action of dopamine and antipsychotic drugs in areas thought to control neuroendocrine function.

The effects of various concentrations of dopamine, norepinephrine and l-isoproterenol on adenylate cyclase activity in a homogenate of the rat median eminence is shown in Figure 1. Low concentrations of dopamine stimulated adenylate cyclase activity. Similar to preparations of the caudate enzyme, l-isoproterenol, a β-agonist, was found to be ineffective on enzyme activity. Norepinephrine was able to stimulate adenylate cyclase activity comparable to dopamine but was not as potent as dopamine at equal doses. The apparent Km for norepinephrine on adenylate cyclase activity of the median eminence was found to be 30 μM whereas that for dopamine agonists apormorphine, N-methyl dopamine and 6,7 dihydroxy-1,2,3,4-tetrahydronapthalene were all found to be potent activators of the adenylate cyclase activity in this brain region.

TABLE 1: CALCULATED INHIBITION CONSTANTS (K_i) FOR ADENYLATE CYCLASE ACTIVITY FROM A HOMOGENATE OF THE RAT MEDIAN EMINENCE

DRUG	K_i* (nM)
Fluphenazine	7.0
Chlorpromazine	70†
Clozapine	61†
Loxapine	13

* *The K_i value was calculated from the relationship $K'_m = 1 + I/Km$ where K'_m and K_m are the concentrations of dopamine required to give half-maximal activation of the enzyme, in the presence and absence of test substance, respectively, and I is the concentration of the inhibitor.*

Where daggers appear, the K_i value was calculated from the relationship $I_{50} = K_i (1 + S/K_m)$, where I_{50} is the concentration of drug required to give 50 per cent inhibition of the enzyme activity, and S is the concentration (40 μM) of dopamine.

Representative drugs from the phenothiazine class were found to be potent inhibitors of the stimulation of adenylate cyclase activity by dopamine (Table 1). Fluphenazine, one of the most potent phenothiazine compounds, both in the treatment of antipsychotic disorders as well as in producing extrapyramidal side effects, was found to be a potent competitive antogonist of dopamine sensitive adenylate cyclase activity in the median eminence. The inhibition constant for fluphenazine on dopamine stimulation of adenylate cyclase activity in a homogenate of the rat median eminence was determined to be 7.0 nM. Another phenothizaine, chlorpromazine, had an inhibition constant ten times higher than that for fluphenazine when tested on the same enzyme preparation. The inhibition constant for chlorpromzaine was calculated to be 70 nM. Clozapine, a dibenzodiazepine, was comparatively equipotent to chlorpromazine. Loxapine, a compound structurally similar to clozapine but much more potent as an antipsychotic and precipitant of extrapyramidal side effects was equivalent to fluphenazine as an inhibitor of dopamine stimulated adenylate cyclase.

The stimulation of adenylate cyclase activity in the median eminence may be especially important from the standpoint of neuroendocrinology. The distribution of dopamine within the median eminence correlates well with the distribution of various releasing hormones in this region including gonadotrophin (LHRH) *(Kizer, Palkivits, Tappaz, Kebabian and Brownstein, 1976)* and dopamine has long been known capable of effecting LH release through an effect on the median eminence *(Kordon, Epelbaum, Enjalbert and McKelvy, 1976)*. Since cyclic AMP has been implicated in the release of hormones from anterior pituitary gland *(Cehovic, Posternak and Charallais, 1971; Bowers, 1971; Zor, Lamprecht, Kaneko, Scheider, McCann, Field, Tsafiri and Lindner, 1971;; Labrie, Borgeat, Lemay, Lemaire, Barden, Drouin, Lemaire, Jolicoeur and Bélanger, 1974; Wakabayaski, Date and Tamaoki, 1973)*, it is possible that dopamine may stimulate the release of hormones from the median eminence which may influence anterior pituitary function. Candidates for such hormones may be LHRH and TRH both of which are localized in this region. The phenothiazines could then block the action of dopamine's effect on such release at the dopamine receptor in the median eminence by blocking the ability of dopamine to activate adenylate cyclase. Certainly, the evidence that a dopamine sensitive adenylate cyclase occurs in the median eminence and that antipsychotic drugs are potent antagonists of this enzyme is compatible with the idea that the endocrinological side effects of the antipsychotic drugs may result from a blockade of dopamine receptors in this area.

Several studies during the past years have implicated the catecholamines in the control of prolactin secretion from the anterior pituitary. Certain pharmacological agents like the phenothiazines have also been implicated in altering prolactin release from the anterior pituitary *(Lu, Amenomori, Chen and Meites, 1970; Ben-David, Dannon, Benveniste, Weller and Sulman, 1971; Smalstig, Swayer and Clemens, 1974)*. While *in vivo* studies suggest that dopamine and the antipsychotic drugs may well have their action on prolactin release at the level of the hypothalamus, specifically at the median eminence, *in vitro* studies indicate that these agents can have a direct action on the pituitary *(Sherman and Kolodny, 1974)*. The results reported herein on dopamine's action on adenylate cyclase activity in the median eminence prompted an investigation of dopamine and phenothiazine action on adenylate cyclase activity in the pituitary. Because of the complexity of the pituitary, it seemed more desirable to conduct such studies on a single cell type. The cells selected for this study were an anterior pituitary clonal cell line known as GH_3/C_{14} which secretes both prolactin and growth hormone.

Cyclic nucleotides have been implicated in prolactin release. Dibutyryl cyclic AMP as well as theophylline have been reported to increase prolactin release *in vitro (Nagasawa and Yanai, 1972; Lemay and Labrie, 1972; Dannies, Gautivik and Tashjian, 1974; Wolff and Jones, 1970)* thus suggesting a positive correlation between the intracellular level of cyclic AMP and the release of prolactin. This was puzzling because dopamine inhibits prolactin release from

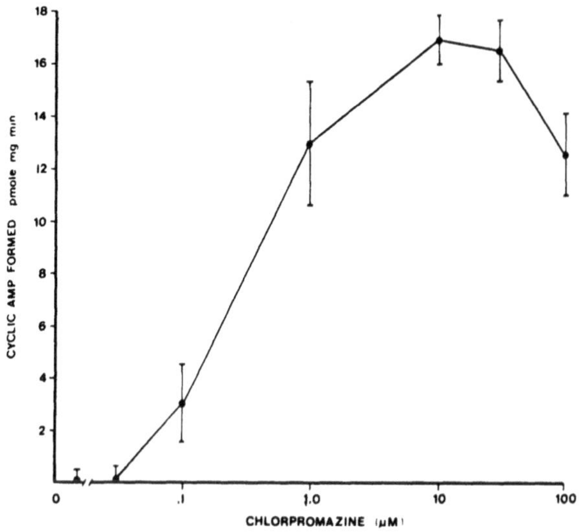

Figure 2. Effect of chlorpromazine on adenylate cyclase activity in a homogenate of GH_3 cells. Standard conditions were used for measurement of adenylyl cyclase activity. The data represent the mean ± SE for six separate experiments.

the anterior pituitary but in the caudate increases cyclic AMP levels. In both the caudate and median eminence, dopamine increases adenylate cyclase activity. Conversely, chlorpromazine, which stimulates prolactin release, had been shown to inhibit dopamine-stimulated responses in the median eminence on cyclase and to block dopamine-stimulated activity on cyclic AMP accumulation in the caudate. Because of the correlation of increased cyclic AMP levels with hormone release, one possibility to consider was that the pituitary might possess an adenylate cyclase activity that could somehow be activated by chlorpromazine. Indeed, such was the case. Homogenates of GH_3/C_{14} cells were found to have an adenylate cyclase system which could be activated by chlorpromazine (Figure 2). The apparent Km in enzyme activity was observed with concentrations of chlorpromazine as low as 0.5 μM. The action of chlorpromazine on adenylate cyclase activity appeared to be specific for prolactin producing cells since the drug had no effect on cyclase activity of AT_{20} cells, a tumor cell line which produces ACTH. Dopamine, either alone or in the presence of chlorpromazine, did not alter the activity of adenylate cyclase in homogenates of the GH_3 cells. The results show that the hydroxy as well as methoxy derivatives mimic the effect of the parent compound on adenylate cyclase activity but were less potent than chlorpromazine itself. The nitrodioxide derivative of chlorpromazine was not effective in stimulating the enzyme. Similarly, 7,8-dihydroxychlorpromazine was virtually ineffective in stimulating GH_3/C_{14} adenylate cyclase activity.

TABLE 2: EFFECT OF CHLORPROMAZINE AND CHLORPROMAZINE DERIVATIVES ON GH_3 ADENYLATE CYCLASE ACTIVITY

DRUG	K_m* ($x10^{-6}M$)
Chlorpromazine	0.7 ± 0.2
Chlorpromazine sulfone	1.8 ± 0.1
7-methoxychlorpromazine	2.5 ± 0.3
7-hydroxychlorpromazine	2.5 ± 0.2
8-hydroxychlorpromazine	2.7 ± 0.4
Chlorpromazine 5-oxide	3.0 ± 0.1
7,8-dimethoxychlorpromazine	3.0 ± 0.1
3,7-dimethoxychlorpromazine	10.0 ± 0.4
Chlorpromazine-N-oxide	30.0 ± 0.2
Chlorpromazine-5, N-dioxide	> 100.00
7,8-dihydroxychlorpromazine	> 100.00

* *Concentration of chlorpromazine and related compounds which cause 50% stimulation of adenylate cyclase activity in a membrane preparation of GH_3 cells.*

The effects of chlorpromazine on adenylate cyclase activity from homogenates of the rat anterior pituitary are shown in Figure 3. Chlorpromazine alone, at concentrations as high as $10^{-4}M$, had no significant effect in these homogenates. However, when chlorpromazine plus the guanyl nucleotide analog, 5'guanylimidophosphate GPP(NH)P, were tested together, a three-fold increase in enzyme activity was observed. A significant increase in cyclase activity occurred in the presence of $1.5 \times 10^{-6}M$ GPP(NH)P. In the presence of $10^{-5}M$ chlorpromazine, a concentration which caused maximal stimulation of the GH_3 enzyme, plus GPP(NH)P, a significant increase in enzyme activity was observed in homogenates of the rat pituitary when compared to GPP(NH)P alone.

These data suggest two sites of action for phenothiazine alteration of adenohypophyseal hormone release. At the level of

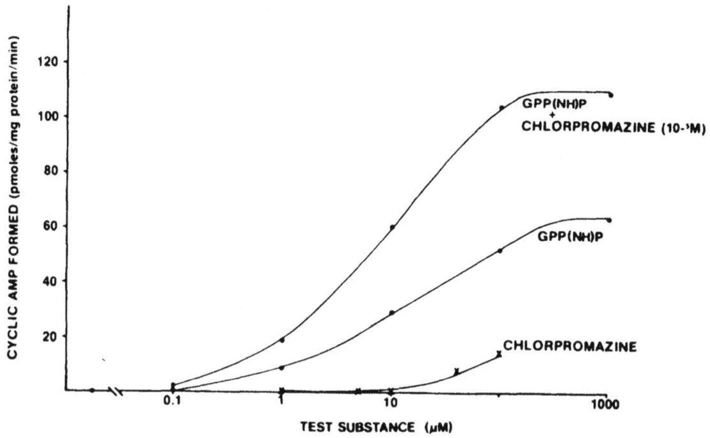

Figure 3. Effect of various concentrations of chlorpromazine, GPP(NH)P or chlorpromazine (10^{-5}M) plus GPP(NH)P on adenylate cyclase activity in homogenates of pituitaries from adult male rats. The increase in cyclic AMP above the basal level is plotted as a function of test substance concentration. In the absence of added test substance 55.6 pmol(mg protein)$^{-1}$ min^{-1} of cyclic AMP was formed. Data are from a representative experiment. Each point is the average of duplicates.

median eminence, the control of hormone release and possible antagonism of this control mechanism may be through the ability of dopamine to stimulate adenylate cyclase activity in the median eminence and the ability of the phenothiazines to block this stimulation. In addition, the mechanism of phenothiazine action on prolactin release may be due to a direct action of the drug on the pituitary, particularly in prolactin-containing cells. Further support for dopamine's role in the functioning of the median eminence comes from the finding that dopamine is capable of binding specifically to homogenates of the rat median eminence and anterior pituitary *(Cronin, Roberts and Weiner, 1978)*. These data suggest that one way the phenothiazines may facilitate increases in prolactin release from the anterior pituitary is through a direct activation of adenylate cyclase in the prolactin-containing cells. Finally, the results of this study indicate that the dopamine receptor in the anterior pituitary, which is apparently uncoupled to adenylate cyclase, has different characteristics from the dopamine receptor cyclase complex in the caudate and median eminence.

ACKNOWLEDGEMENTS

Supported in part by grants from the Pharmaceutical Manufacturer's Association and Hoffman La Roche.

REFERENCES

Anden, N.-E. and Stock, G. (1973). Effect of clozapine on the turnover of dopamine in the corpus striatum and in the limbic system. *J. Pharm.Pharmac.* *25*, 346-348.

Ben-David, N., Dannon, A., Benveniste, R., Weller, C.P. and Sulman F.G. (1971). Results of radioimmunoassays of rat pituitary and serum prolactin after adrenalectomy and perphenazine treatment in rats. *J. Endocr.* *50*, 599-606.

Birge, C.A.,Jacobs, L.S., Hammer, C.T. and Daughaday, W.H. (1970). Catecholamine inhibition of prolactin secretion by isolated rat adenohypophyses. *Endocrinology* *86*, 120-130.

Bowers, C.Y. (1971). Studies on the role of cyclic AMP in the release of anterior pituitary hormones. *Ann. N.Y. Acad. Sci.* *185*, 52,-540.

Carlsson, A. and Lindqvist, M. (1963). Effect of chloropromazine or haloperidol on formation of 3-methoxytyramine and normetanephrine in mouse brain. *Acta pharmac. tox.* *20*, 140-144.

Cehovic, G., Posternak, T. and Charallais, E. (1971). A study of biological activity and resistance to phosphodiesterase of some derivatives and analogues of cyclic AMP. *Adv. Cyclic Nucleo. Res.* *1*, 521-540.

Clement-Cormier, Y.C., Kebabian, J.W., Petzold, G.L. and Greengard, P. (1974). Dopamine-sensitive adenylate cyclase in mammalian brain: A possible site of action of antipsychotic drugs. *Proc. natn. Acad. Sci. U.S.A.* *71*, 1113-1117.

Cronin, M.J., Roberts, J.M. and Weiner, R.I. (1978). Dopamine and dihydroergocryptine binding to the anterior pituitary and other brain areas of the rat and sheep. *Endocrinology* (in press).

Dannies, P., Gautivik, K. and Tashjian, A. (1974). Thyrotropin releasing hormone-cyclic AMP interrelationships in prolactin-producing cells in culture. *Endocr. Soc. Abs.* *56*, A-151.

Hollister, L.E. (1972). *Clinical Use of Psychotherapeutic Drugs*, volume 1. Springfield, Illinois: Charles C. Thomas.

Hornykiewicz, O. (1973). Parkinson's disease: from brain homogenate to treatment. *Fed. Proc.* *32*, No. 2, 183-190.

Iversen, L.L. (1975). Dopamine receptors in the brain a dopamine-sensitive adenylate cyclase models synaptic receptors, illuminating antipsychotic drug action. *Science* *188*, 1084-1089.

Kamberi, I.A., Mical, R. and Porter, J.C. (1971). Effect of anterior pituitary perfusion and intraventricular injection of catecholamines on prolactin release. *Endocrinology* *88*, 1012-1020.

Kavanagh, A. and Weisz, J. (1973/74). Localization of dopamine and norepinephrine in the medial basal hypothalamus of the rat. *Neuroendocrinology* *13*, 201-212.

Kebabian, J.W., Petzold, G.L. and Greengard, P. (1972). Dopamine-sensitive adenylate cyclase in caudate nucleus of rat brain, and its similarity to the "Dopamine Receptor". *Proc. natn.*

Acad. Sci. U.S.A. 69, 2145-2149.

Kizer, J.S., Palkovits, M., Tappaz, M., Kebabian, J. and Brownstein, M.J. (1976). Distribution of releasing factors, biogenic amines and related enzymes in the bovine median eminence. *Endocrinology 98,* 685-695.

Koch, Y., Lu, H. and Meites, J. (1970). Biphasic effects of catecholamines on pituitary prolactin release *in vitro. Endocrinology 87,* 673-675.

Kordon, D., Epelbaum, J., Enjalbert, A. and McKelvy, J. (1976). Neurotransmitter interactions with neuroendocrine tissue. In: *Subcellular Mechanisms in Reproductive Neuroendocrinology* pp. 167-184. Eds. F. Naftalin, K.J. Ryan and I.J. Davies. Elsevier, New York.

Labrie, F., Borgeat, O., Lemay, A., Lemaire, S., Barden, N., Drouin, J., Lemaire, I., Jolicoeur, P. and Bélanger, A. (1974). Role of cyclic AMP in the action of hypothalamic regulatory hormones. *Adv. Cyclic Nucleo. Res. 5,* 787-801.

Lemay, A. and Labrie, F. (1972). Calcium-dependent stimulation of prolactin release in rat anterior pituitary *in vitro* by N^6-monobutyryl adenosine 3',5'-monophosphate. *FEBS Letters 20,* 7-10.

Lu, K.-H., Amenomori, Y., Chen, C.-L. and Meites, J. (1970). Effects of central acting drugs on serum and pituitary prolactin levels in rats. *Endocrinology 87,* 667-672.

MacLeod, R.M. (1969). Influence of norepinephrine and catecholamine-depleting agents on the synthesis and release of prolactin and growth hormone. *Endocrinology 85,* 916-923.

Martin, J.B. (1973). Neural regulation of growth hormone secretion. *New Engl. J. Med. 288,* 1384-1392.

Nagasawa, H. and Yanai, R. (1972). Promotion of pituitary prolactin release in rats by dibutyryl adenosine 3',5'-monophosphate. *J. Endocr. 55,* 215-216.

Nybäck, H. and Sedvall, G. (1968). Effect of chloropromazine on accumulation and disappearance of catecholamines formed from tyrosine C^{14} in brain. *J. Pharmac. exp. Ther. 162,* 294-301.

Quijada, M., Illner, P., Krulich, L. and McCann, S.M. (1974). The effects of catacholamines on hormone release from anterior pituitaries and ventral hypothalami incubated *in vitro. Neuroendocrinology 13,* 151-163.

Shaar, C.J. and Clemens, J.A. (1974). The role of catecholamines in the release of anterior pituitary prolactin *in vitro. Endocrinology 95,* 1202-1212.

Sherman, L. and Kolodny, H.D. (1974). The effects of drugs on human hypophysiotropic functions. In: *Mammary Cancer and Neuroendocrine Therapy.* pp. 369-400. Ed. B.A. Stall. Butterworths, London.

Smalstig, E.B., Swayer, B.D. and Clemens, J.A. (1974). Inhibition of rat prolactin release by apomorphine *in vivo* and *in vitro. Endocrinology 95,* 123-129.

Wakabayaski, K., Date, Y. and Tamaoki, B. (1973). On the mechanism of action of luteinizing hormone-releasing factor and prolactin release inhibiting factor. *Endocrinology 92*, 698-704.

Wolff, J. and Jones, A.B. (1970). Inhibition of hormone-sensitive adenyl cyclase by phenothiazines. *Proc. natn. Acad. Sci. 65*, 454-459.

BIOGENIC AMINE-STIMULATED ADENYLATE CYCLASE AND SPIROPERIDOL-BINDING SITES IN RABBIT BRAIN: EVIDENCE FOR SELECTIVE LOSS OF RECEPTORS WITH AGING

M.H. MAKMAN,[1,2] H.S. AHN,[1] L.J. THAL,[3] B. DVORKIN,[1]
S.G. HOROWITZ,[1] N.S. SHARPLESS,[4] and M. ROSENFELD[2]

Departments of Biochemistry[1], Molecular Pharmacology[2],
Neurology[3] and Psychiatry[4]
Albert Einstein College of Medicine
New York, New York

INTRODUCTION

The aging process in man results in or is accompanied by a host of changes in brain structure and biochemistry *(Terry and Gershon, 1976)*. The possible relationship of these alterations to the diminished mental function seen frequently in the older population is not yet known. Furthermore, the implications of these alterations for the manifestation and severity of neurological diseases such as parkinsonism are poorly understood. A major task in the investigation of biochemical changes occurring with senescence is to sort out the early or primary changes of greatest functional significance. In this regard, it is becoming increasingly evident that biogenic amine and closely interrelated transmitter or synaptic modulator systems play important roles in hypothalamic, cortical and extrapyramidal system functions. A number of studies have indicated that in experimental animals and in man deficiencies in brain aminergic systems, assessed functionally and/or biochemically, may occur with senescence *(Finch, 1973; Finch, Jonec, Hody, Walker, Morton-Smith, Alper and Dougher, 1975; Jonec and Finch, 1975; McGeer and McGeer, 1976ab; McGeer, McGeer and Suzuki, 1977; Simpkins, Mueller, Huang and Meites, 1977)* (see also Finch and McGeer in this volume). An overall conclusion that may be derived from these biochemical studies is that with senescence decreased levels and/or capacity for synthesis of transmitters such as dopamine, acetylcholine and gamma-aminobutyric acid (GABA) may occur in selective brain regions.

Another and major component of transmitter systems that as yet has received little attention in the overall evaluation of transmitter action in the senescent brain is the receptor for the transmitter in question. Receptors might possibly be altered in number, properties and/or functional coupling. Furthermore, changes might possibly occur in pre-synaptic as well as in post-synaptic receptor systems. Central aminergic receptor systems have been studied biochemically primarily by either direct assessment of radioligand binding to receptors *(Synder and Bennett, 1976)* or measurement of receptor-mediated stimulation of adenylate cyclase activity by transmitter *(Makman, 1977)*. Methodology for study of direct binding of biogenic amine agonists or antagonists to receptors has developed relatively recently. Suitable and relatively specific ligands are now available for study of dopamine, α-adrenergic and β-adrenergic receptors in various brain regions, but not for histamine receptors and less clearly for serotonin receptors. A variety of intact and broken cell systems have been utilized to demonstrate neurotransmitter modulation of cyclic AMP formation as well as the possible role of cyclic AMP in synaptic transmission and other cellular functions in the nervous system. Both intact brain slices and tissue homogenates have been useful for study of regulation of cyclic AMP formation by transmitters.

Homogenates have been particularly useful for study of dopamine-stimulated adenylate cyclases *(Brown and Makman, 1972; Kebabian, Petzold and Greengard, 1972; Iversen, 1975; Makman, 1977)*. The regional distribution of dopamine receptors in the central nervous system, assessed by dopamine-stimulation of adenylate cyclase *(Makman, 1977)* and by both *in vitro* and *in vivo* binding of radioactive dopamine receptor antagonists to membranes *(Creese, Burt and Snyder, 1975; Burt, Creese and Snyder, 1976; Fields, Reisine and Yamamura, 1978; Laduron, Janssen and Leysen, 1978; Höllt, Czlonkowski and Herz, 1977)* agrees with that predicted on the basis of anatomical, behavioral, and electrophysiologic studies. Studies of the relative affinities of agonists and antagonists for striatal dopamine-stimulated adenylate cyclases and for displacement of labeled dopamine antagonists or agonists from receptor sites in striatum strongly suggest that even in a given brain region there may be more than one type or form of dopamine receptor. Based on binding studies, it appears that dopamine receptors may possibly exist in both agonist or antagonist states *(Burt et al., 1976)*. Neither of these receptor forms in striatum, however, is clearly identical to that which is coupled to adenylate cyclase *(Makman, 1977)*. In studies of monkey brain regions, we have distinguished pharmacologically three types of dopamine receptors coupled to adenylate cyclase, one type present in anterior limbic cortex, a second in frontal cortex and a third in several regions including caudate nucleus, retina and hypothalamus *(Mishra, Makman, Ahn, Dvorkin, Horowitz, Keehn and Demirjian, 1976; Ahn, Mishra, Demirjian and Makman, 1976; Ahn and Makman, 1977)*. Dopamine receptor-adenylate cyclase systems in rat striatum and cortex are also distinguishable from one another *(Brockaert, Tassin, Thierry, Glowinski and Premont, 1977)*.

The studies to be presented here are concerned primarily with the influence of aging on the dopamine-stimulated adenylate cyclase system and on dopamine antogonist binding in various regions of the rabbit central nervous system. Certain related parameters are evaluated as well. In addition, studies are presented concerning histamine-stimulated adenylate cyclases. Finally, preliminary data for a cortical β-adrenergic adenylate cyclase are also presented. Particular attention has been given to brain regions most likely to be involved in central functions known or thought to be lost or impaired with senescence in man. The results indicate that selective changes in brain aminergic receptor function do occur with senescence. Also, on the basis of this work it appears that the rabbit provides a new and useful animal model for elucidation of changes in brain transmitter systems with aging.

METHODS

Rabbits (New Zealand White, from Breathitt Rabbit Producers, Jackson, Kentucky) were sacrificed by air injection into an ear vein followed by opening of the cranial vault and rapid dissection of the brain. Aged rabbits were retired breeder females (average mean life span of 5 years) and for comparison female rabbits 4-5 months of age and animlas of both sexes at 2-3 weeks of age were used. The average mean life span of male rabbits from the same commercial breeder is 7 years. Male Sprague-Dawley rats were killed by decapitation.

For adenylate cyclase assays,aliquots of the homogenate (25 μl) were incubated in a shaking water bath at 30°C for 2.5 or 5 min. in a total volume of 100 μl of incubation medium containing 80 mM Tris maleate buffer (pH 7.4), 5 mM theophylline, 2 mM $MgSO_4$, 0.5 mM ATP and appropriate test agents. The reaction was terminated by placing the assay tubes in a boiling water bath. Particulate matter was removed by low speed centrifugation and aliquots of the supernatant fluids were assayed for cyclic AMP by a protein binding assay previously described *(Brown and Makman, 1973)*. Each experiment involved, as a minimum, triplicate adenylate cyclase incubations for every condition studied and replicate determinations of cyclic AMP content were made in most cases. Other details and analyses of the data were as reported previously *(Ahn and Makman, 1977)*. Assays of dopamine receptor binding sites in rabbit brain membrane fractions (resuspended 50,000 x g pellets) were carried out according to the method of Burt *(et al., 1976)* with the following modifications. The ligand used was $[H^3]$-spiroperidol, 26 Ci/mmole (Amersham Searle) at concentrations of 0.1 to 1.6 nM. The final volume for the incubations was 1.0 ml. Incubations were carried out for 20 rather than 10 minutes. Specific binding was measured as the excess over the blank value obtained in the presence of 1 μM spiroperidol. The blank produced essentially the same value as 1 μM (+)-butaclamol and the pharmacologically inactive isomer (-)-butaclamol did not displace radioactive ligand at the concentration chosen for the blank. Saturability of stereospecific binding sites was demonstrated in all brain regions studied.

Choline acetylase activity of tissue homogenates was determined as described previously *(Datta, Thal and Wajda, 1971)*. This procedure measures the rate of acetylation of choline by C^{14}-acetyl CoA in the presence of excess substrate, with the C^{14}-acetylcholine formed then separated from unreacted C^{14}-acetyl CoA by electrophoresis.

Dopamine and norepinephrine were determined in rabbit brain regions by a radioenzymatic method previously described *(Sharpless and Brown, 1978)*.

RESULTS

TRANSMITTER-STIMULATED ADENYLATE CYCLASES IN REGIONS OF RABBIT CNS

Dopamine was found to simulate adenylate cyclase activity not only in rabbit striatum (caudate-putamen) and retina as previously reported *(Makman, Brown and Mishra, 1975)* but also in frontal cortex, anterior limbic (cingulate) cortex and hypothalamus (Table 1). The data presented are for a maximally effective concentration of dopamine. The stimulatory effects are comparable in magnitude to those obtained with these regions from other species studied with the exception of monkey which,in general,has CNS activity more responsive to dopamine. The responses to dopamine in the rabbit CNS regions have been characterized in other studies, not presented

TABLE 1. INFLUENCE OF BIOGENIC AMINES ON ADENYLATE CYCLASE ACTIVITY IN VARIOUS REGIONS OF YOUNG ADULT RABBIT BRAIN AND RETINA

Region	Adenylate cyclase activity (pmoles cyclic AMP formed/mg protein /5 min)*			
	Basal	Dopamine	Histamine	Norepinephrine
Anterior limbic cortex	246 ± 23(5)	494 ± 25(4)	514 ± 22(4)	----
Frontal cortex	237 ± 18(5)	462 ± 17(5)	574 ± 33(5)	481 ± 43(2)
Hypothalamus	480 ± 46(2)	715 ± 24(2)	725 ± 38(2)	701 ± 62(2)
Caudate-Putamen	301 ± 52(6)	763 ± 67(6)	----	----
Retina	183 ± 34(6)	458 ± 21(6)	----	----

* Each experimental value represents the mean ± S.E.M. for separate experiments with tissues from 4-6 month old female rabbits (number of rabbits and experimenta in parentheses). The final concentration of agents used was 100 μM.

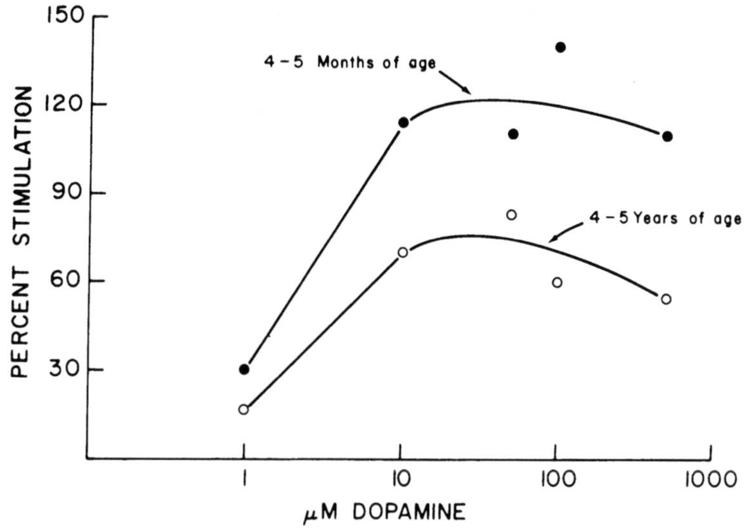

Figure 1. Influence of age on dopamine-stimulated adenylate cyclase of rabbit caudate-putamen. Experimental details were as described in the text. Values represent percent stimulation by dopamine at the indicated concentrations over basal activity as given in Table 2. Values are means for separate assays of tissue from 7 young and 7 old animals (young and old animals paired for study).

here, to involve interactions with dopamine receptors based on the relative potencies of agonists, dopamine, apomorphine and norepinephrine, and on the blockade of those reponses by dopamine antagonists such as fluphenazine, haloperidol and pimozide. In other species,the major component of the response to norepinephrine in the regions shown (Table 1) appears to be due also to interaction with dopamine receptors *(Ahn et al., 1976; Ahn and Makman, 1977)*. Further studies will be required to establish this in the rabbit. Rabbit frontal cortex also contains,in a relatively low amount,a β-adrenergic receptor adenylate cyclase system stimulated by isoproterenol (Figure 2).

The two rabbit cortical regions studied and also rabbit hypothalamus contain histamine-stimulated adenylate cyclase activity in addition to activity stimulated by dopamine (Table 1). Spiker, Palmer and Manian *(1976)* have also studied this activity in rabbit cortex. We find histamine-stimulated activity to be present also in frontal cortex of *Cebus* monkey and guinea pig *(Ahn and Makman, unpublished observations)* and in hypothalamus

TABLE 2. LACK OF INFLUENCE OF AGING ON BASAL AND Gpp (NH)p-STIMULATED ADENYLATE CYCLASE ACTIVITY OF RABBIT CNS REGIONS.

Region and additions to assay	Activity* 4-5 month old animals	Activity* 5 year old animals	Percent change in activity with age
Striatum (caudate-putamen):			
Basal	368 ± 58 (7)	334 ± 32 (7)	-9
Anterior limbic cortex:			
Basal	246 ± 23 (4)	234 ± 27 (4)	-5
Frontal cortex:			
Basal	237 ± 18 (5)	249 ± (4)	+5
Gpp (NH)p, 100 μM	5550 ± 125 (2)	5539 ± 304 (2)	-4
Hypothalamus:			
Basal	480 ± 46 (2)	490 ± 28 (2)	+2
Gpp (NH)p, 100 μM	1618 ± 48 (2)	1563 ± 235 (2)	-3
Retina:			
Basal	183 ± 34 (6)	189 ± 22 (6)	+3

* pmoles cyclic AMP formed/5 min/mg protein (number of animals, assayed seperately in triplicate, in parentheses).

of guinea pig *(Ahn and Makman, 1977)*. Although not yet characterized pharmacologically in the rabbit cortex, our preliminary studies of monkey and guinea pig cortex indicate that a mixed H1 and H2 receptor type is involved in the stimulation by histamine. While we presume the dopamine-stimulated adenylate cyclase to be neuronal, it is possible that a major component of histamine-stimulated adenylate cyclase in at least some brain regions may be glial or vascular (see also Discussion).

INFLUENCE OF AGING ON BASAL AND Gpp(NH)p-STIMULATED ADENYLATE CYCLASE ACTIVITY

No change in basal adenylate cyclase activity was found with aging in any of the regions of brain or in retina of the rabbit (Table 2). Basal activity was also unchanged in rat striatum as these animals aged from 2-3 months to 2-2.5 years of age (data not presented). Similarly, in frontal cortex and hypothalamus of the rabbit no alteration was found with aging in the marked stimulation of adenylate cyclase activity by the GTP analogue, Gpp(NH)p (Table 2).

TABLE 3. ADENYLATE CYCLASE ACTIVITY OF HYPOTHALAMUS OF YOUNG AND OLD RABBITS

Additions to Assay	Increase in cyclic AMP formation* due to biogenic amine		% change in response with age †
	4-5 month old animals	5 year old animals	
Dopamine, 5 μM	106 ± 10	22 ± 12	-79 (p<0.001)
Dopamine, 100 μM	235 ± 24	161 ± 10	-31 (P<0.02)
Norepinephrine, 100 μM	220 ± 62	74 ± 25	-66 (p<0.05)
Histamine, 100 μM	245 ± 38	83 ± 15	-66 (p<0.005)

* p moles cyclic AMP/5min/mg protein formed over basal activities as given in Table 2.

† % change in stimulation due to biogenic amine with age (p values for significance of change in parentheses).

INFLUENCE OF AGING ON DOPAMINE-STIMULATED ADENYLATE CYCLASE ACTIVITY OF RABBIT AND RAT STRIATUM

While basal adenylate cyclase activity of striatum (caudate-putamen) was unchanged with aging, in the older rabbits the maximal stimulation of activity by dopamine was decreased by approximately 40-50 percent (Figure 1). There was no apparent change in the concentration of dopamine required for half-maximal stimulation. Similar decreases with age in maximal stimulation by dopamine were

TABLE 4. ADENYLATE CYCLASE ACTIVITY OF ANTERIOR LIMBIC CORTEX OF YOUNG AND OLD RABBITS

Additions to assay	Increase in cyclic AMP formation* due to biogenic amine		% change in response with age †
	4-5 month old animals	5 year old animals	
Dopamine, 100 μM	248 ± 24 (+101%)	129 ± 23 (+55%)	-46 (p<0.005)
Histamine, 5 μM	194 ± 32 (+ 79%)	117 ± 12 (+50%)	-37 (p<0.05
Histamine, 100 μM	268 ± 22 (+109%)	164 ± 28 (+70%)	-36 (p<0.01)

* pmoles cyclic AMP/5min/mg protein formed over basal activities .e given in Table 2 (data for 4 separate experiments with young and 4 with old animals). (Percent stimulation by biogenic amine in parentheses).

† % change in stimulation due to biogenic amine with age (p values for significance of change in parentheses).

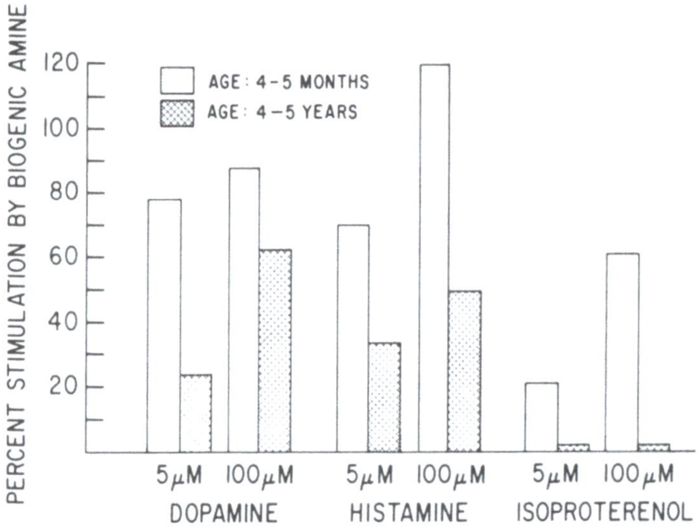

Figure 2. Influence of age on biogenic amine-stimulated adenylate cyclase activity of rabbit frontal cortex. Experimental details were as described in the text. Values represent percent stimulation over basal activity as given in Table 2.

also obtained for rat striatum (as animals aged from 2-3 months to 2-2.5 years of age, data not shown).

INFLUENCE OF AGING ON BIOGENIC AMINE-STIMULATED ADENYLATE CYCLASES OF RABBIT HYPOTHALAMUS, CORTICAL AREAS AND RETINA

In hypothalamus, both dopamine- and histamine-stimulated adenylate cyclase activities were decreased in the older animals (Table 3). Similar results were obtained for stimulation by dopamine and by histamine in anterior limbic cortex (Table 4) and frontal cortex (Figure 2). Finally, in preliminary studies there was also found a markedly decreased stimulation by isoproterenol in frontal cortex of the older animals (Figure 2).

These studies indicated that the effects of aging on biogenic amine-stimulated adenylate cyclases were selective for the transmitter or hormonal component of the system (i.e., did not involve basal or Gpp(NH)p-stimulated activity) but were not selective for a particular transmitter system or for one particular brain region. On the other hand, in at least one CNS region containing appreciable

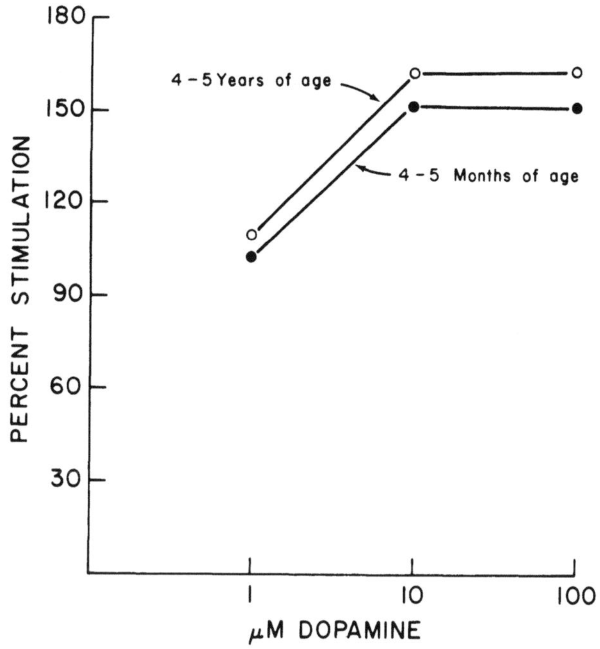

Figure 3. Influence of age on dopamine-stimulated adenylate cyclase activity of rabbit retina. Experimental details were as described in the text. Values represent percent stimulation over basal activity as given in Table 2.

amounts of dopamine-stimulated adenylate cyclase, the retina, there was found to be no change in this activity with aging (Figure 3). Thus, the retina, in a sense, serves as a control for the other regions and provides evidence for at least some degree of regional selectivity in the influence of aging on dopamine-stimulated activity.

ADENYLATE CYCLASE ACTIVITY OF CNS REGIONS OF DEVELOPING RABBITS

Previously, we reported that in rat retina there occurred a rapid postnatal development of dopamine-stimulated adenylate cyclase during the first few weeks after birth during which period development and maturation of the inner retina (where the activity appears to be located) also occurred *(Makman et al., 1975)*. Adenylate cyclase activity of CNS regions of 2 to 3 week old rabbits is summarized in Table 5. As evident from those data together with the data in Tables 1 and 2 for 4 to 6 month old animals, basal and transmitter-stimulated adenylate cyclases were fully developed not only in retina but also in frontal cortex and

TABLE 5. ADENYLATE CYCLASE ACTIVITY OF CNS REGIONS OF 2-3 WEEK OLD RABBITS

Region and addition to the Assay	Activity* †	Percent Stimulation by transmitter †
Frontal Cortex		
None (basal)	200 ± 19	
Gpp(NH)p, 100 μM	1105 ± 282	
Dopamine, 5 μM	380 ± 44	90 ± 22
Dopamine, 100 μM	541 ± 42	171 ± 21
Histamine, 100 μM	397 ± 50	99 ± 25
Hypothalamus		
None (basal)	162 ± 5	
Gpp(NH)p, 100 μM	873 ± 50	
Dopamine, 100 μM	248 ± 18	53 ± 11
Histamine, 100 μM	279 ± 13	72 ± 8
Caudate-Putamen		
None (basal)	448 ± 22	
Dopamine, 0.1 μM	670 ± 42	66 ± 24
Dopamine, 1 μM	928 ± 25	107 ± 29
Dopamine, 10 μM	1046 ± 40	134 ± 10
Dopamine, 100 μM	1234 ± 57	175 ± 14
Retina		
None (basal)	146 ± 26	
Dopamine, 1 μM	362 ± 18	134 ± 12
Dopamine, 10 μM	378 ± 14	145 ± 10
Dopamine, 100 μM	379 ± 20	146 ± 15

* pmoles cyclic AMP formed/5 min/mg protein

† Data represent averages for 3 experiments ± S.E.M. for caudate-putamen and retina or for triplicate assay incubation ± S.E.M. for 1 experiment for frontal cortex and hypothalamus

caudate-putamen at 2 to 3 weeks of age. There occurred in these regions no appreciable change in basal or transmitter-stimulated activities with further maturation to the young adult (4 to 6 month old) animal. On the other hand, basal activity in hypothalamus and Gpp(NH)p-stimulated activity in both hypothalamus and frontal cortex *increased* during this maturational period but not as animals aged from 4 to 6 months to 5 years as described earlier. These studies indicate that the changes that were found to occur with

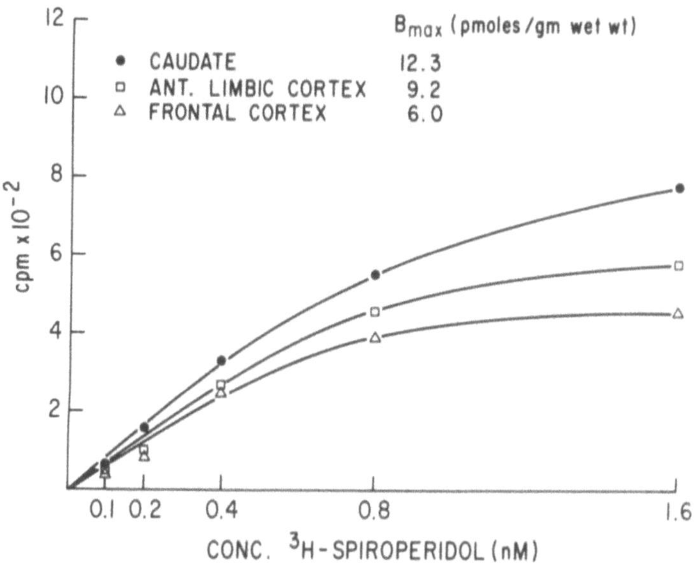

Figure 4. Representative saturation curves for specific [^3H]-spiroperidol binding sites in rabbit brain regions. Experimental conditions were as described in the text. Data were obtained from a single old rabbit.

senescence were not due to the continuation of a process already evident during an earlier period of life but represent rather phenomena with post-maturity onset.

INFLUENCE OF AGING ON SPIROPERIDOL-BINDING SITES IN RABBIT BRAIN REGIONS

Based on the studies of adenylate cyclase, in which the maximal response to transmitter (e.g., dopamine) in several brain regions was selectively decreased with senescence, it appeared that this might be due directly to a decreased number of receptors. This possibility was assessed directly by measurement of binding of radioligand to dopamine receptors. With the ligand chosen, the dopamine antagonist, spiroperidol, saturable, stereospecific binding sites could be measured not only in caudate-putamen, but in cortical regions as well. Representative data for saturation curves in three brain regions from a single rabbit are shown in

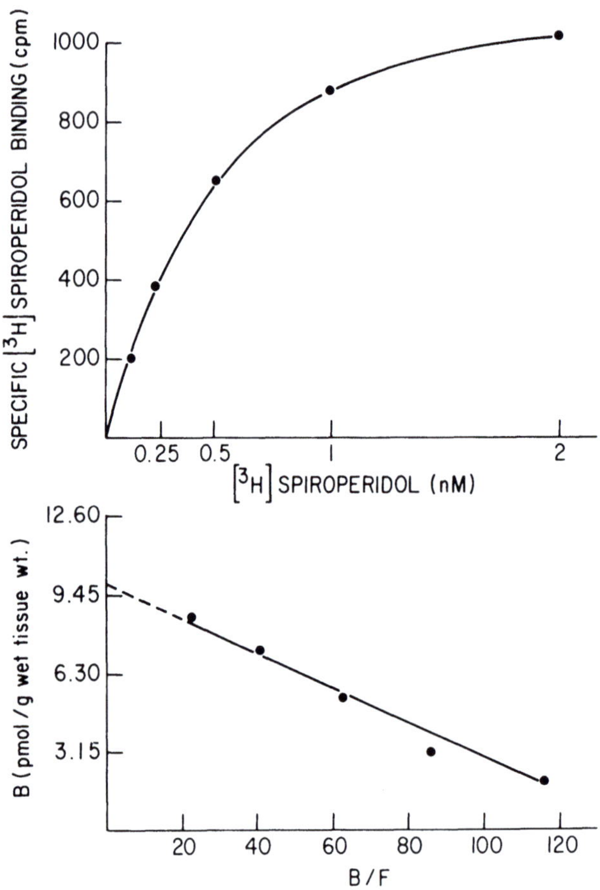

Figure 5. Representative saturation curve (upper portion) and Scatchard analysis derived from that curve (lower portion) for specific [^{3}H]-spiroperidol binding sites in rabbit caudate-putamen. Experimental conditions were as described in the text. Data were obtained from a single young rabbit.

Figure 4. The binding data were evaluated by Scatchard analysis, as exemplified by Figure 5. The results of these binding studies in young and old animals indicated an appreciable decrease in number of spiroperidol binding sites in caudate-putamen with aging (Table 6). This decrease was comparable in magnitude to that described earlier for dopamine-stimulated adenylate cyclase of caudate-putamen. Similar decreases in spiroperidol binding sites

TABLE 6. [^3H]-SPIROPERIDOL BINDING SITES IN BRAIN REGIONS OF OLD AND YOUNG RABBITS

	Caudate-Putamen	Frontal Cortex	Anterior limbic cortex
Total binding sites (fmoles/mg wet weight:			
4-5 month old rabbits*	10.3 (8)	4.1 (8)	7.1 (9)
5 year old rabbits*	6.7 (6)	3.8 (6)	4.3 (7)
Ratio of binding sites in young/old rabbits +	1.8 ± 0.13 (6)	1.74 ± 0.64 (4)	1.65 ± 0.55 (4)

* Number of rabbits in parentheses.

† Values ± S.E.M. are given (number of pairs of old and young rabbits assayed simultaneous in parentheses). Ratios are computed for the paired data.

were found with aging in frontal and anterior limbic cortex (Table 6). However, because of the greater variation in binding capacity from one animal to another in those cortical regions, further studies will be required to establish the significance of these differences. It should be noted that the ratios of binding sites in young and old rabbits in Table 6 are for paired animals (assayed at the same time) only, whereas the average values for binding sites in each region as given in Table 6 include data from additional (non-paired) animals.

In contrast to the decreased number of spiroperidol binding sites in the older animals, the affinity of the receptor for spiroperidol was not changed with age in any of the regions studied. Thus, in caudate-putamen K_D values for spiroperidol ± SEM (number of experiments in parentheses) were 1.4 ± 0.36 (7) and 0.97 ± 0.23 (6) nanomolar for young and old rabbits respectively. Corresponding values in anterior limbic cortex were 1.1 ± 0.21 (7) and 1.1 ± 0.32 (6).

INFLUENCE OF AGING ON DOPAMINE CONTENT AND CHOLINE ACETYLASE ACTIVITY

Of major interest with respect to the changes with aging found to occur in receptors for biogenic amines, was to determine whether this involved a selective loss of receptors on a neuron that was still viable and capable of other functions, or the selective loss of this neuron with retention of other neurons (e.g., presynaptic neuron, etc.), or loss of both pre- and post-synaptic neurons. To approach this problem, we have carried out analyses of catecholamine concentration and also activity of the enzyme choline acetylase in these brain regions. The data obtained revealed that in rabbit caudate-putamen no change in dopamine concentration occurred with aging from 4 to 5 months to 5 years of age (data not

TABLE 7. LACK OF INFLUENCE OF AGE ON CHOLINE ACETYLASE ACTIVITY OF RABBIT BRAIN

Brain Region	Young (5 mos)	Activity* Old (5 yrs)	Ratio young/old
Caudate-Putamen	3.6	5.5	0.66
Frontal Cortex	0.92	1.01	0.91
Anterior limbic cortex	0.91	1.05	0.87

* μmoles acetyl choline formed/hr/gm wet wt.

shown). Since dopamine content may be considered an indirect measure of the presynaptic dopamine nerve terminals in caudate-putamen, it appears that no loss of the presynaptic component is evident at an age at which a major decrease is evident in the associated postsynaptic receptor component. Studies are currently in progress to assess the influence of aging in transmitter content in cortex and hypothalamus of the rabbit.

Since the major role of dopamine in caudate-putamen is believed to be the inhibition of firing of cholinergic neurons in that region, we also investigated the activity of choline acetylase in caudate-putamen (Table 7). No decrease in the activity of this enzyme was found with age. Thus, the loss of dopamine receptors was not accompanied by loss of an enzyme of transmitter synthesis located in neurons presumed to also contain a major component of the post-synaptic dopamine receptors. In addition, there occurred with senescence no change in choline acetylase activity in the two cortical regions studied (Table 7).

DISCUSSION

The studies presented here indicate that in the rabbit the amount of dopamine-stimulated adenylate cyclase activity and the number of spiroperidol binding sites in caudate-putamen both decrease with senescence. We have obtained similar findings for rat striatal dopamine stimulated adenylate cyclase. Decreased activity in striatum of aged rats has also been reported by Puri and Volicer *(1977)*. Furthermore, Finch and coworkers have found a decreased number of spiroperidol binding sites in mouse striatum with senescence, as reported elsewhere in this volume. Thus, in three species striatal dopamine receptors assessed by adenylate cyclase activity and/or spiroperidol binding capacity have been found to be decreased with aging. That the loss of striatal dopamine receptors with age is selective was indicated by the lack of change in basal adenylate cyclase activity, dopamine concentration or activity of choline acetylase in rabbit striatum. It, therefore, seems likely that the process by which dopamine receptors are decreased in this brain region takes place in the absence of, or at least is initiated prior to, comparable loss of either the presynaptic dopaminergic input or of the neurons on which are located the dopamine receptors. Based on the less complete but similar data obtained here for frontal cortex, anterior limbic cortex and hypothalamus of the rabbit, comparable losses of dopamine receptors, presumably due to a similar mechanism, also occurred in these other regions with senescence. On the other hand, in the retina no loss of dopamine receptors coupled to adenylate cyclase was evident.

The actual mechanism of dopamine receptor loss is not yet known. We have previously suggested that central dopamine receptors may be modulated either in the direction of supersensitivity or desensitization by physiological or pathological processes *(Makman, 1977)*. Thus, unilateral lesions of the dopamine neurons in substantia nigra of the rat resulted in enhanced stimulation by dopamine of adenylate cyclase in the striatum *(Mishra, Gardner, Katzman and Makman, 1974)*. Also, during normal postnatal development in the rat the dopamine-stimulated adenylate cyclase of the rat after attaining maximal activity by 15 days of age, subsequently declines in activity from day 15 to 29 *(Makman et al., 1975)*. The reason for this subsequent decline (to adult levels) was postulated to be related to a desensitization occurring shortly after innervation, i.e., the reverse of denervation supersensitivity. Recently, Dr. Ahn has carried out studies demonstrating the direct production of desensitization of dopamine receptors by incubation of monkey (data not shown) or rabbit (Table 8) cortical slices *in vitro* with

TABLE 8. INFLUENCE OF PREINCUBATION OF RABBIT FRONTAL CORTEX SLICES WITH DOPAMINE ON SUBSEQUENT RESPONSIVENESS OF HOMOGENATE ADENYLATE CYCLASE TO DOPAMINE AND HISTAMINE

Agents added to assay	30 minute preincubation*: without dopamine	with dopamine (100 μM)	P values
	Adenylate cyclase activity**		
None (basal)	158 ± 14	208 ± 26	NS
Dopamine, 100 μM	244 ± 15 (+54)	201 ± 2 (-3)	p<0.05
Histamine, 100 μM	301 ± 16 (+91)	272 ± 15 (+31)	NS

* Chopped tissue slices of frontal cortex from 4-5 month old female rabbits were preincubated for 30 min. at 37°C in Krebs-Ringer Bicarbonate medium containing 15 mM glucose either in the presence or absence of 100 μM dopamine. At the end of the incubation the slices were washed three times with the same medium and then homogenized in the standard Tris-maleate-EGTA buffer for adenylate cyclase assay

** pmoles cyclic AMO formed/mg protein/5 min. Values are means ± S.E.M. for triplicate incubations and assays for each condition studied.

dopamine. The slice incubation is then followed by extensive washing of the tissue and homogenization for assay of adenylate cyclase activity. As shown in the table, selective desensitization to dopamine was produced in such an experiment. We have also been able to produce selective desensitization to histamine by preincubation of tissue with histamine *(Ahn and Makman, unpublished observations)*. Kakiuchi and Rall *(1968)* first demonstrated desensitization of brain tissue to norepinephrine and histamine in studies carried out entirely with slices of rabbit cerebellum, rather than preincubation of slices followed by adenylate cyclase assay. Studies made by our laboratory first showed that a comparable desensitization of β-adrenergic receptors coupled to adenylate cyclase was produced by exposure of intact lymphoid cells or cultured fibroblasts to catecholamines *(Makman, 1971)*. Other studies of desensitization in cultured cells have been reviewed previously *(Makman, Morris and Ahn, 1977)*.

Thus, there is a long-standing precedent for the acute desensitization of both central and peripheral tissues to biogenic amines. The involvement of such a mechanism in the changes with aging reported here would also provide a common process by which loss of responsiveness not only to dopamine but also to histamine and β-adrenergic agonists might occur. Desensitization might be responsible for loss of peripheral receptor responsiveness with aging as well, e.g., for the reduced β-receptors found on human mononuclear cell membranes with aging *(Schocken and Roth, 1977)*. Different tissues or cell types in culture normally differ in the rapidity and/or extent of the desensitization process. In general, dopamine receptors appear to be desensitized much less readily than are β-receptors. The changes with aging might be related to changes in sensitivity of receptors to the desensitization process. Alternatively, the recovery from desensitization might be involved. It is, of course, also possible that there is a selective decline in the rate of synthesis of receptors occurring without specific relationship to the desensitization process. Additional studies will be required to establish the actual mechanism of receptor loss.

The studies reported here indicate a similar rate of loss of different types of dopamine receptors with age. Thus, cortical dopamine receptors and striatal dopamine receptors coupled to adenylate cyclase may be distinguished from one another pharmacologically as discussed previously, and both are decreased with age in the rabbit. Also, spiroperidol binding sites represent a class of dopamine receptors, at least in part, distinguishable from those measured by the adenylate cyclase assay, and in several regions both classes of receptors are decreased with senescence.

In addition, the receptor losses reported in this study suggest the involvement of several different cell types. While we believe the dopamine receptors to be exclusively neuronal in the rabbit brain, histamine and β-adrenergic receptors may be located in part or largely on glial and/or vascular cells, and also the extent to which these receptors may be neuronal or non-neuronal may very in different brain regions *(Wilkening and Makman, 1975)*.

The functional significance of the receptor losses found with aging remain to be established. The similar decreases in dopamine receptor binding and dopamine-stimulated adenylate cyclase at least suggest the absence of spare dopamine receptors coupled or potentially coupled to adenylate cyclase. Following nearly complete nigral lesions, we routinely observe about a doubling of striatal dopamine-stimulated adenylate cyclase activity in the rat *(Mishra et al., 1974)*. Smaller increases (20-40%) in haloperidol binding sites are observed in striatum after nigral lesion *(Creese et al., 1977)*. It is presumed that these changes make a significant contribution to the functional receptor supersensitivity which occurs after such a lesion. In order to produce readily detectable behavioral effects by destruction of nigral dopamine neurons, lesions generally must be at least 80 percent complete (assessed by extent of loss of striatal dopamine content). *(E. Gardner, unpublished observations)*. Thus, at the presynaptic level there may exist greater "reserves" than are present at the postsynaptic level in striatum. The "postsynaptic" dopamine receptor losses reported here therefore might well have functional significance. While symmetrical bilateral loss might have little apparent influence on basal motor function, such loss might impair function in specific situations or be additive to loss occurring for some other reason. In man, such losses with senescence in striatum might possibly result in predisposition to development of parkinsonism. Likewise, losses in cortex and hypothalamus might result in an impairment of other CNS functions with age.

SUMMARY

Evidence for selective decreases in biogenic amine receptor function with age in the rabbit has been obtained. Dopamine-stimulated adenylate cyclase activity in the striatum (caudate-putamen) of rabbit brain declined by about 50 percent as rabbits aged from less than 1 to 4 to 5 years of age. Similar decreases in transmitter-stimulated adenylate cyclase activity were found for histamine as well as for dopamine and norepinephrine in hypothalamus, frontal cortex and anterior limbic cortex. Isoproterenol-stimulated activity was also decreased with age in frontal cortex. These changes appeared to represent decreases in maximal response and not alteration in affinity for amine. In contrast, dopamine-stimulated adenylate cyclase of retina and transmitter-independent (basal or Gpp(NH)p-stimulated) activity in each of the regions studied were not altered with age. Dopamine receptors in striatum directly assessed by measurement of [^{3}H]-spiroperidol binding revealed a comparable decrease in the number of binding sites without change in ligand affinity. Preliminary data also indicated decreased spiroperidol binding sites in the cortical regions of older animals. These changes in striatum and cortex were evident in the absence of decreases in either dopamine content or choline acetylase activity, an activity presumed to be present in neurons containing dopamine receptors. It is proposed that selective age-dependent decreases in postsynaptic biogenic amine receptor content occur in the absence of, or independent from, neuronal cell loss, possibly by a mechanism involving receptor desensitization. These changes occur in the animal model in those brain regions which in man are thought to be of importance in the loss of cerebral function that is found with senscence.

A preliminary report of this work has be presented *(Makman, Ahn, Thal, Dvorkin, Horowitz, Sharpless and Rosenfeld, 1978)*.

ACKNOWLEDGEMENTS

This work was supported by NIH research grants AG00374 and NS09649. The authors wish to thank Dr. Mohammed Kalimi for suggesting to us the use of the rabbit brain as an aging model and for his helpful comments during the course of these studies.

REFERENCES

Ahn, H.S. and Makman, M.H. (1977). Neurotransmitter-sensitive adenylate cyclase in the hypothalami of guinea-pig, rat and monkey. *Brain Res. 138*,125-138.

Ahn, H.S., Mishra, R.K., Demirjian, C. and Makman, M.H. (1976). Cathecholamine-sensitive adenylate cyclase in frontal cortex of primate brain. *Brain Res. 116*, 437-454.

Brockaert, J., Tassin, J.P., Thierry, A.M., Glowinski, J. and Premont, J. (1977). Characteristics of dopamine and β-adrenergic sensitive adenylate cyclases in the frontal cerebral cortex of the rat. Comparative effects of neuroleptics on frontal cortex and striatal dopamine sensitive adenylate cyclases. *Brain Res. 122*, 71-86.

Brown, J.H. and Makman, M.H. (1972). Stimulation by dopamine of adenylate cyclase in retinal homogenates and of adenosine-3'5' cyclic AMP formation in intact retina. *Proc. natn. Acad. Sci. 69*, 539-543.

Brown, J.H. and Makman, M.H. (1973). Influence of neuroleptic drugs and apomorphine on dopamine-sensitive adenylate cyclase of retina. *J. Neurochem. 21*, 477-479.

Burt, D.R., Creese, I. and Snyder, S.J. (1976). Properties of [^{3}H]haloperidol and [^{3}H] dopamine binding associated with dopamine receptors in calf brain membranes. *Mol. Pharmac. 12*, 800-812.

Creese, I., Burt, D.R. and Snyder, S.H. (1975). Dopamine receptor binding differentiation of agonist and antagonist states with ^{3}H-dopamine and ^{3}H-haloperidol. *Life Sci. 17*, 993-1002.

Creese, I., Burt, D.R. and Snyder, S.H. (1977). Dopamine receptor binding enhancement accompanies lesion-induced behavioral supersensitivity. *Science 197*, 596-598.

Datta, K., Thal, L. and Wajda, I. (1971). Effects of morphine on choline acetyltransferase levels in the caudate nucleus of the rat. *Br. J. Pharmac. 41*, 84-93.

Fields, J.Z., Reisine, T.D. and Yamamura, H.I. (1978). Biochemical demonstration of dopaminergic receptors in rat and human brain using ^{3}H-spiroperidol. *Brain Res.* (In Press).

Finch, C.E. (1973). Catecholamine metabolism in the brains of aging male mice. *Brain Res. 52*, 261-276.

Finch, C.E., Jonec, V., Hody, G., Walker, J.P., Morton-Smith, W., Alper, A. and Dougher, G.J., Jr. (1975). Aging and the passage of L-tyrosine, L-DOPA, and insulin into mouse brain slices *in vitro*. *J. Geront. 30*, 33-40.

Höllt, V., Czlonkowski, A. and Herz, A. (1977). The demonstration *in vivo* of specific binding sites for neuroleptic drugs in mouse brain. *Brain Res. 130*, 176-183.

Iversen, L. (1975). Dopamine receptors in the brain: a dopamine-sensitive adenylate cyclase models synaptic receptors, illuminating antipsychotic drug action. *Science 188*, 1084-1089.

Jonec, V.J. and Finch, C.E. (1975). Aging and dopamine uptake by subcellular fractions of the C57BL/6J male mouse brain. *Brain Res. 91*, 197-215.

Kakiuchi, S. and Rall, T.W. (1968). The influence of chemical agents on the accumulation of adenosine 3,'5'-phosphate in slices of rabbit cerebellum. *Mol. Pharmac. 4*, 367-378.

Kebabian, J.W., Petzold, G.L. and Greengard, P. (1972). Dopamine-sensitive adenylate cyclase in caudate nucleus of rat brain and its similarity to the "dopamine receptor". *Proc. natn. Acad. Sci. 69*, 2145-2149.

Laduron, P.M., Janssen, P.F.M. and Leysen, J. (1978). Spiperone: a ligand of choice for neuroleptic receptors. 2. Regional distribution and *in vivo* displacement of neuroleptic drugs. *Biochem. Pharmac. 27*, 317-321.

McGeer, P.L. and McGeer, E.G. (1976a). Enzymes associated with the metabolism of catecholamines, acetylcholine and GABA in human controls and patients with Parkinson's disease and Huntington's chorea. *J. Neurochem. 26*, 65-76.

McGeer, E.G. and McGeer, P.L. (1976b). In: *Neurobiology of Aging, vol. 3*, pp. 389-403. Eds. R. Terry and S. Gershon. Academic Press, New York.

McGeer, P.L., McGeer, E.G. and Suzuki, J.S., (1977). Aging and extrapyramidal function. *Arch. Neurol. 34*, 33-35.

Makman, M.H. (1971). Properties of adenylate cyclase of lymphoid cells. *Proc. natn. Acad. Sci. 68*, 885-889.

Makman, M.H. (1977). Actions of cyclic AMP and its relationship to transmitter function in nervous tissue. In: *Biochemical Actions of Hormones, vol 4*, pp. 407-496. Ed. G. Litwack. Academic Press, New York.

Makman, M.H., Brown, J.H. and Mishra, R.K. (1975). Cyclic AMP in retina and caudate nucleus: influence of dopamine and other agents. *Adv. Cyc. Nucleo. Res. 5*, 661-679.

Makman, J.H., Morris, S.A. and Ahn, H.S. (1977). Cyclic nucleotides. In: *Growth, Nutrition and Metabolism of Cells in Culture, vol. 3*, pp. 295-354. Eds. G.H. Rothblat and V.J. Cristofalo. Adacemic Press, New York.

Makman, M.H., Ahn, H.S., Thal, L., Dvorkin, B., Horowitz, S.G., Sharpless, N. and Rosenfeld, M. (1978). Decreased brain biogenic amine-stimulated adenylate cyclase and and spiro-peridol-binding sites with aging. *Fed. Proc. 37*, 548.

Mishra, R.K., Gardner, E.L., Katzman, R. and Makman, M.H. (1974). Enhancement of dopamine-stimulated adenylate cyclase activity in rat caudate after lesions in substantia nigra: evidence for denervation supersensitivity. *Proc. natn. Acad. Sci. 71*, 3883-3887.

Mishra, R.K., Makman, M.H., Ahn, H.S., Dvorkin, B., Horowitz, S.G., Keehn, E. and Demirjian, C. (1976). Differences in primate brain regions in relative potency for antagonism of dopamine-stimulated adenylate cyclases by neuroleptic drugs and possible implications for localization of antipsychotic activity. *Neurosci. Abst. 2*, 1134.

Puri, S.K. and Volicer, L. (1977). Effect of aging on cyclic AMP levels and adenylate cyclase and phosphodiesterase activities in the rat corpus striatum. *Mech. Aging. Dev. 6*, 53-58.

Schocken, D.D. and Roth, G.S. (1977). Reduced β-adrenergic receptor concentrations in aging man. *Nature 267*, 856-858.

Sharpless, N.S. and Brown, L.L. (1978). Use of microwave irradiation to prevent post-mortem catecholamine metabolism: evidence for tissue disruption artifact in a discrete region of rat brain. *Brain Res. 140*, 171-176.

Simpkins, J.W., Mueller, G.P., Huang, H.H. and Meites, J. (1977). Evidence for depressed catecholamine and enhanced serotonin metabolism in aging male rats: possible relation to gonado-tropin secretion. *Endocrinology 100*, 1672-1678.

Snyder, S.H. and Bennett, J.P., (1976). Neurotransmitter receptors in the brain: biochemical identification. *A. Rev. Physiol. 38*, 153-175.

Spiker, M.D., Palmer, G.C. and Manian, A.A. (1976). Action of neuroleptic agents on histamine-sensitive adenylate cyclase in rabbit cerebral cortex. *Brain Res. 104*, 401-406.

Terry, R.D. and Gershon, S. (1976). *Neurobiology of Aging*. Academic Press, New York.

Wilkening, D. and Makman, M.H. (1975). 2-Chloroadenosine-dependent elevation of adenosine 3',5'-cyclic monophosphate levels in rat caudate nucleus slices. *Brain Res. 92*, 522-528.

RELATIVE QUANTITATION OF MONOAMINE HISTOFLUORESCENCE IN YOUNG AND OLD NON-HUMAN PRIMATES

JOHN R. SLADEK, Jr. and CELIA D. SLADEK

Departments of Anatomy and Neurology
University of Rochester School of Medicine
Rochester, New York 14642

ABSTRACT

The relative content of monoamines within identified brain stem neurons in 4 and 20-year old monkeys *(Macaca nemestrina)* was determined with microspectrofluorometric techniques. Intraneuronal monoamine content was found to decrease with age in the locus coeruleus, substantia nigra and raphe and was found to be reduced further in the locus coeruleus by the presence of lipofuscin granules. These data indicate that intraneuronal monoamine content is reduced in the aged macaque.

INTRODUCTION

Aging is accompanied by decreases in content and turnover of some central monoamines as well as changes in levels of their synthetic enzymes *(Finch, 1973; this volume; Carlsson, this volume; Clemens, this volume)*. Clearly, the loss of dopamine in the nigrostriatal and mesolimbic systems and its relationship to parkinsonian symptoms is well appreciated. Significant loss of dopamine in the hypothalamus of the aged rodent may be related to reproductive senescence, and the possibility exists that other brain monoamine systems also change with age.

Histofluorescence analysis of monoaminergic neurons in the aged animal model can allow the analysis of qualitative and quantitative changes in pools of neurons as well as within individual neurons. This type of analysis can provide information concerning: (1) whether all neurons of a specific neucleus change synchronously with age; (2) the degree to which aging pigments may affect intraneuronal monoamine content; and (3) the relative amount of intraneuronal monoamine within different brain nuclei. The present investigation represents an attempt to determine the relative changes in intraneuronal content of dopamine, norepinephrine and serotonin which accompany age in a non-human primate.

MATERIALS AND METHODS

Six colony-raised, pit-tailed Macaques *(Macaca nemestrina)* aged 4 and 20 years old were examined with the formaldehyde-induced histofluorescence technique of Falck and Hillarp *(1962)*. Relative quantitation of monoamine histofluorescence was determined microspectrofluorometrically as described elsewhere *(Sladek, McNeill, Walker and Sladek, in press)*. All spectral recordings were performed at constant instrument settings to eliminate variation in instrument sensitivity from interfering with quantitative values. The locus coeruleus, substantia nigra, and nucleus raphe dorsalis were chosen as representative nuclei for norepinephrine, dopamine and serotonin-containing neurons respectively. The mean and standard deviation of all quantitations were calculated

RESULTS

Qualitatively, histofluorescence of each of the nuclear groups chosen for analysis appeared most intense in the 4-year old animals, although perikarya of the substantia nigra appeared less intense than either the locus coeruleus or raphe in both ages. Serotonin fluorescence of raphe neurons appeared within the perinuclear cytoplasm and even extended into neuronal processes in the young animals (Figures 1 and 2). These same neurons occasionally possessed lipofuscin granules, and in the old animals lipofuscin fluorescence predominated. Perikarya of the substantia nigra often appeared to contain a reticulated cytoplasm,and in general, the cells looked less healthy than those of the other two groups (Figures 3 and 4). Lipofuscin was not a common feature of nigral neurons. Norepinephrine fluorescence of the locus coeruleus appeared intense in both ages, somewhat more intense in the younger animals (Figures 5 and 6). Lipofuscin granules were seen in locus coeruleus neurons in both ages; however, they were more abundant in the older animals.

Spectral analysis of the relative intensity of individual neurons of these groups revealed a significant loss of monoamine from each group with age. The relative intensities of dopamine-containing neurons of the substantia nigra were analyzed as three groups of 4 and 20-year old animals. In each group the relative intensity was significantly lower in the older animals (Figure 7). Similar determinations were made for serotonin-containing neurons of the nucleus raphe dorsalis and norepinephrine-containing neurons of the locus coeruleus and confirmed the qualitative analysis that monoamine fluorescence was less intense in the older animals (Figures 8 and 9). Furthermore, locus coeruleus was subdivided according to the presence or absence of lipofuscin granules within the perinuclear cytoplasm. In both ages, the mean norepinephrine

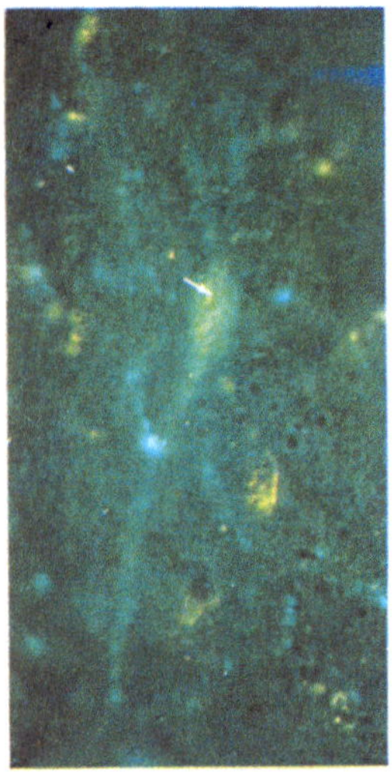

Fig. 1

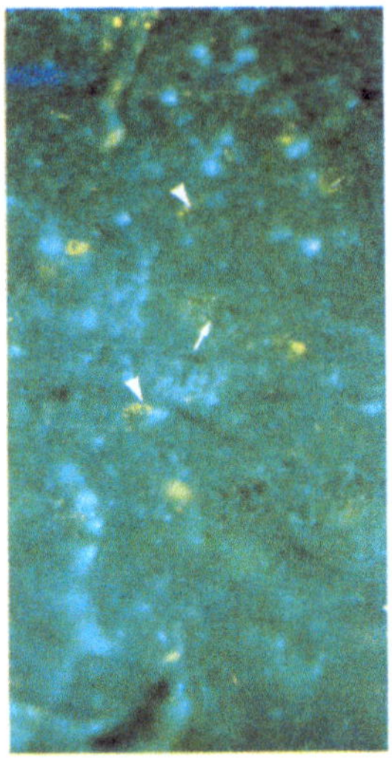

Fig. 2

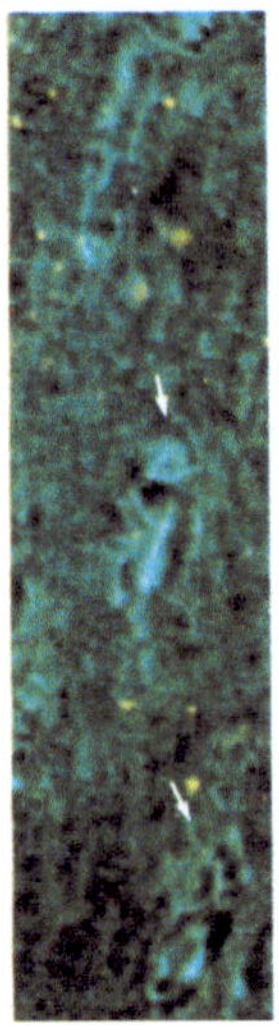

Fig. 3

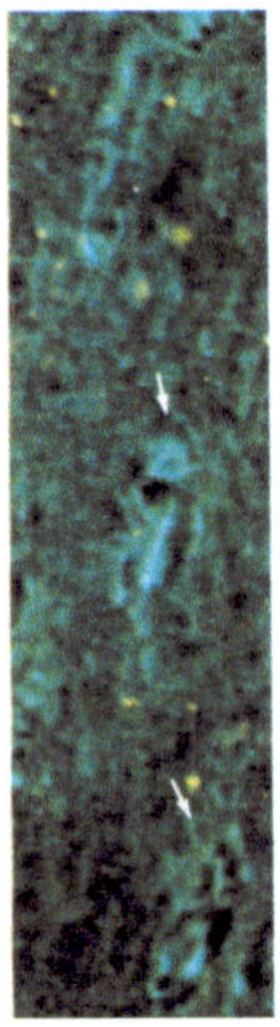

Fig. 4

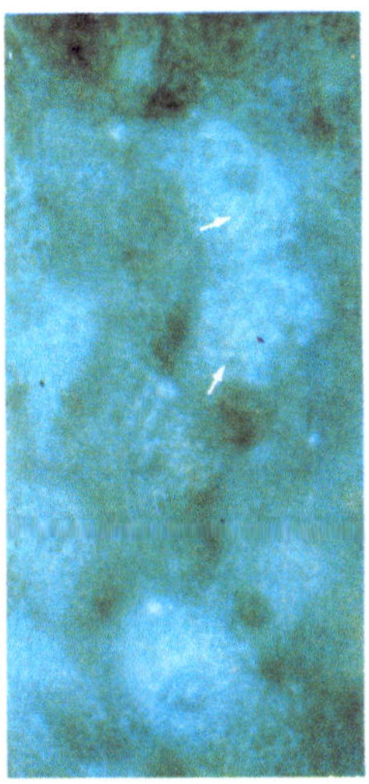

Fig. 5

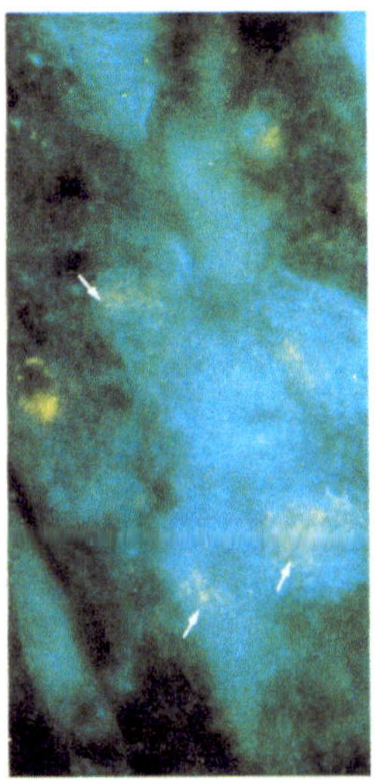

Fig. 6

Figure 1. *(top left) Nucleus raphe dorsalis - 4-year old. Yellow serotonin histofluorescence appears in the cytoplasm and process of a neuronal perikaryon. Yellow-gold pigment granules (→) also are seen in the cytoplasm. X200.*

Figure 2. *(top right) Nucleus raphe dorsalis - 20-year old. A predominance of lipofuscin granules is seen (►) and serotonin histofluorescence (→) appears less prominent than that depicted in Figure 1. X200.*

Figure 3. *(middle left) Substantia nigra - 4-year old. Moderately intense dopamine histofluorescence is seen within perikarya (→). X200.*

Figure 4. *(middle right) Substantia nigra - 20 year-old. Neuronal perikarya (→) appear to contain a weaker histofluorescence and a more reticulated cytoplasm than younger animals. X200.*

Figure 5. *(bottom left) Locus coeruleus - 4-year old macaque. Intense blue norepinephrine fluorescence characterized the perinuclear cytoplasm. Yellow-gold pigment granules (→) are visible in some of the neurons. X200.*

Figure 6. *(bottom right) Locus coeruleus - 20-year old macaque. Norepinephrine fluorescence is somewhat less intense and pigment granules (→) are more prominent in comparison to the young macaque. X400.*

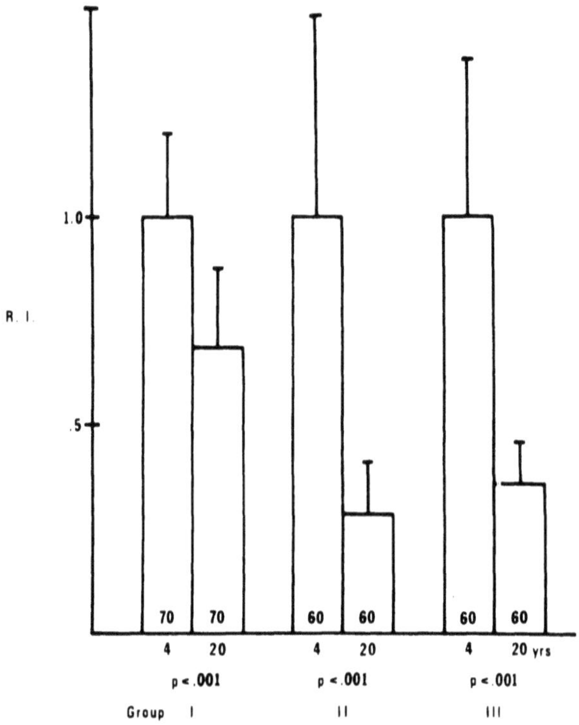

Figure 7. Histogram of relative intensity (RI) for substantia nigra. The mean and standard deviation are plotted for 3 pairs of 4 and 20-year old macaques. The number of perikarya analyzed appears within each bar above the horizontal axis. Significant decreases in neurotransmitter content appeared in the 20-year old macaques.

intensity was significantly less in the lipofuscin-positive perikarya than in the lipofuscin-free perikarya (Figure 10). Also, the mean intensity of the lipofuscin-free neurons of the 20-year old was significantly less than the 4-year old, and the intensity of lipofuscin-positive neurons was significantly lower in the 20-year old. When all the fluorescent neurons, both lipofuscin-free and positive, were pooled and treated as a single group, it was found that the mean intensity of these neurons in the 20-year old was significantly less than the 4-year old.

DISCUSSION

Analysis of selected monoamine-containing loci of the aged non-human primate revealed a decrease in fluorescence intensity accompanying aging. While this is a well-known phenomenon in the

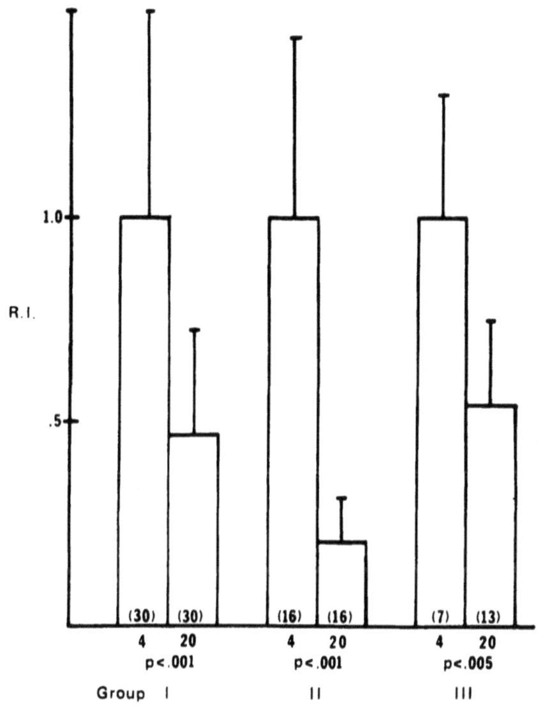

Figure 8. Histogram of relative intensity (RI) for nucleus raphe dorsalis. The mean and standard deviation are plotted for 3 pairs of 4 and 20-year old macaques. The number of perikarya analyzed appears within each bar above the horizontal axis. Significant decreases in neurotransmitter content appeared in the 20-year old macaques.

nigrostriatal dopamine system, it has not been demonstrated previously in any of the other monoamine groups of the brain. However, as Professor Carlsson has alluded to in his presentation, it seems reasonable that similar changes might occur in other nuclei especially in light of the reported alterations in monoamine synthetic enzymes which accompany human aging *(McGeer and McGeer, 1975)*. It is not known at present if all monoamine systems lose neurotransmitter content with age, but preliminary examination in aged macaque of the ventral noradrenergic system including its terminal innervation in the hypothalamus revealed an intence histofluorescence comparable to that seen in young macaque *(Sladek et al., in press)*. Furthermore, the tuberoinfundibular dopamine system of the 20-year old macaque demonstrated maximum intensity, perhaps even above that of the younger animals *(Sladek et al., in press)*.

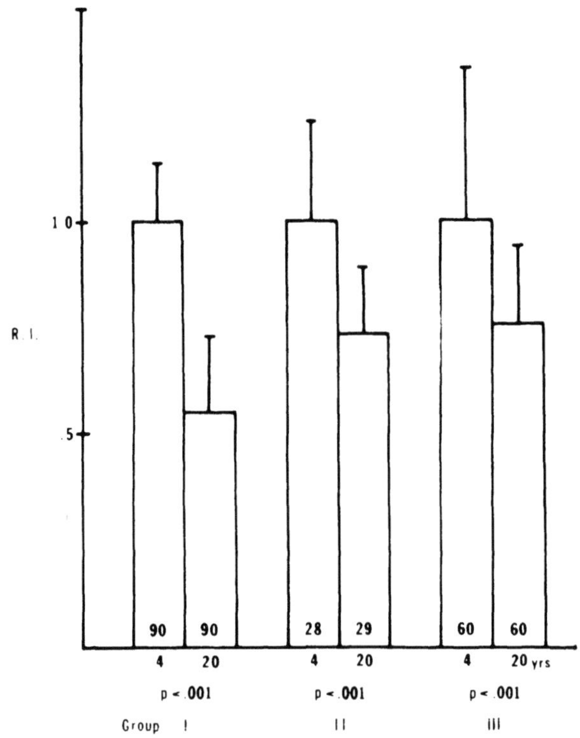

Figure 9. Histogram of relative intensity (RI) for locus coeruleus. The mean and standard deviation are plotted for 3 pairs of 4 and 20-year old macaques. The number of perikarya analyzed appears within each bar above the horizontal axis. Significant decreases in neurotransmitter content appeared in the 20-year old macaques.

Thus, a drop in neurotransmitter content in neurons of monoaminergic loci may not be a generalized phenomenon, but this needs to be tested further.

The occurrence of lipofuscin granules within neurons during aging is a well-known phenomenon *(Brizzee, Ordy and Kaack, 1974)*. The present data indicate that in the locus coeruleus the presence of lipofuscin within the perinuclear cytoplasm is associated with decreased monoamine neurotransmitter content. To our knowledge, this is the first demonstration that these autofluorescent pigment deposits are coincidental with decreased neurotransmitter content. This raises speculation about possible depressed transmitter function in monoaminergic neurons which contain lipofuscin. Alternately, the decreased content could be due to a displacement of norepinephrine by pigment granules, however, spectral analysis of that part of the cytoplasm which contained only norepinephrine in

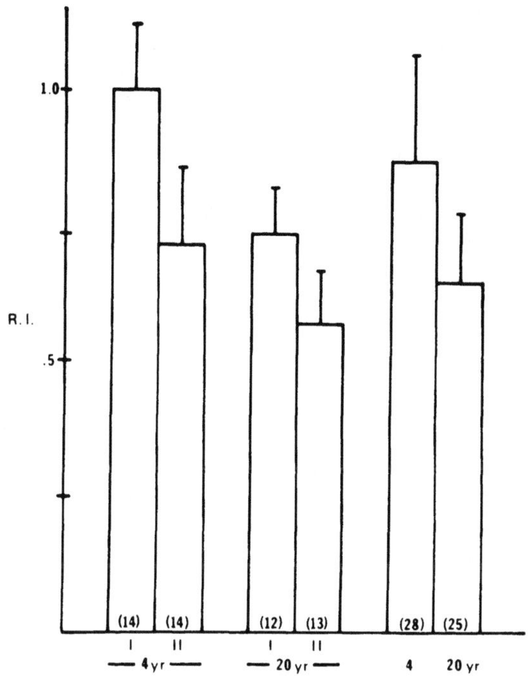

Figure 10. The effect of autofluorescent pigment granules on norepinephrine relative intensity is illustrated for neurons of the locus coeruleus. In a pair of 4 and 20-year old macaques, locus coeruleus neurons were subdivided into lipofuscin-free (group I) and lipofuscin-bearing (group II) perikarya. The presence of lipofuscin correlated with a significant decrease in norepinephrine content in both the 4 and 20-year old macaques ($p < .001$). Furthermore, norepinephrine content was significantly lower ($p < .001$) in lipofuscin-free perikarya of the 20-year old macaque in comparison to lipofuscin-free perikarya of the 4-year old. The mean intensity of lipofuscin-positive neurons also was significantly lower in the 20-year old animal than in the 4-year old. ($p < .001$). Mean ± SD.

lipofuscin-bearing perikarya revealed a significant loss of norepinephrine. Relative intensity values were all measured at the peak emission wavelength of norepinephrine (i.e., 480 nm). It is unlikely that the presence of lipofuscin in the scanned neurons augmented this peak, for the lipofuscin emission peaks (560, 600 nm) are considerably higher than that of norepinephrine. Furthermore, if an error had been introduced by this procedure, it would have enhanced the relative intensity values of the lipofuscin-bearing scans. Instead, these values always appeared less than those gathered from lipofuscin-free neurons. Therefore, the present results appear to indicate that the presence of lipofuscin within coeruleal neurons is coincidental with a depressed transmitter content in comparison to lipofuscin-free coeruleal neurons. The presence of lipofuscin would not appear to be the only factor related to decreased neurotransmitter content because lipofuscin-free perikarya of the locus coeruleus in the 20-year old animal contained significantly less norepinephrine than lipofuscin-free perikarya of the 4-year old animal. Possible causes for this reduced content of neurotransmitter could include increased turnover, decreased synthesis or decreased re-uptake. The elucidation of this problem will await further investigation.

ACKNOWLEDGEMENTS

The authors thank Betsy Putnam Buell for technical assistance and Dr. Douglas Bowden who supplied the animals for this investigation. This work was supported in part by NIH Grant NS 11642 and by USPHS Career Development Award NS-00259 to Celia D. Sladek.

REFERENCES

Brizzee, K.R., Ordy, J.M. and Kaack, B. (1974). Early appearance and regional differences in intraneuronal and extra neuronal lipofuscin accumulation with age in the brain of a non-human primate (Maccaca mulatta) *J. Geront. 29*, 366-381.

Carlsson, A. Age-dependent changes in central dopaminergic and other monoaminergic systems. In: *Parkinson's Disease: Aging and Neuroendocrine Relationships.* Eds. C.E. Finch, D.E. Potter and A.D. Kenny. Plenum Press. New York (in press).

Clemens, J. Some neuroendocrine aspects of aging. In: *Parkinson's Disease: Aging and Neuroendocrine Relationships.* Eds. C.E. Finch, D.E. Potter and A.D. Kenny. Plenum Press. New York (in press).

Falck, B., Hillarp, N.S., Thieme, G. and Torp, A. (1962). Fluorescence of catecholamines and related compounds condensed with formaldehyde. *J. Histochem. Cytochem. 10*, 348-354.

Finch, C.E. (1973). Catecholamine metabolism in the brains of aging male mice. *Brain Res. 52*, 261-276.

Finch, C.E. Age-related changes in brain catecholamines: A symposium of findings in C57BL/6J mice and other rodent models. In: *Parkinson's Disease: Aging and Neuroendocrine Relationships.* Eds. C.E. Finch, D.E. Potter and A.D. Kenny. Plenum Press. New York (in press).

McGeer, E.G. and McGeer, P.L. (1975). Age changes for some enzymes associated with metabolism of catecholamines, GABA and acetylcholine. *Adv. Behav. Biol. 16*, 287-305.

Sladek, J.R., Jr., McNeill, T.H., Walker, P. and Sladek, C.D. Age-related alterations in monoamine and neurophysin systems in primate brain. In: *Aging in Non-Human Primates, Neurobiology of Aging, Vol. 6*. Ed. D.M. Bowden, Raven Press (in press).

INTEGRATED MORPHOLOGY OF NEURONAL CATECHOLAMINES AND NEUROPHYSIN IN THE AGED MACAQUE

JOHN R. SLADEK, Jr., JOANN McCONNEL,* and THOMAS H. McNEILL

Department of Anatomy
University of Rochester School of Medicine
Rochester, New York

and Department of Neurobiology and Anatomy*
University of Texas Medical School
Houston, Texas

ABSTRACT

A new method for the simultaneous visualization of brain peptides and monoamine neurotransmitters was employed to analyze the integrated morphology of hypothalamic catecholamines and neurophysins in young and old monkeys. Immunocytochemical analysis using bovine neurophysin revealed a dual papulation of light and dark stained cells in the paraventricular nucleus in young and old macaques. In general, both populations of neurons stained with less density in old macaques indicating the possibility of a reduced content of neurophysin. Further analysis using specific neurophysin antisera for vasopressin or oxytocin revealed an appreciable decrease in the number of vassopressin-containing perikarya in the 20 year monkey whereas oxytocin-synthesizing neurons did not show a similar change in numbers with age. Qualitatively, terminal innervation patterns of hypothalamic catecholamines remained strikingly constant in spite of marked reductions in dark-stained neurophysin perikarya of the paraventricular nucleus.

INTRODUCTION

A number of brain functions are regulated or at least influenced by neuron systems which either are located within or connected to the hypothalamus. Some of these intrinsic neuronal networks have been identified as catecholaminergic or peptidergic through the independent use of immunocytochemical and histofluorescence techniques. In many instances, components of peptidergic and catecholaminergic (CA) neurons have been found in the same neural locus; notable examples include: (1) norepinephrine terminals and vasopressin (VP) and oxytocin (OX) neurons of magnocellular hypothalamic nuclei *(Fuxe, 1965; Zimmerman, Defendini, Sokol and Robinson, 1975)*; (2) endocrine-releasing hormones and dopamine-containing terminals of the median eminence *(Hökfelt, Elde, Fuxe, Johansson, Ljungdahl, Goldstein, Luft, Efendic, Nilsson, Terenius, Ganten, Jeffcoate, Rehfeld, Said, Perez de la Mora, Possani, Tapia, Teran and Palacios, 1978)*. Independent analyses of the magnocellular nuclei have led to considerable speculation that pericellularly arranged noradrenergic terminals of the supraoptic (SON) and paraventricular (PVN) nuclei may contact peptidergic perikarya providing a morphological basis for a functional interaction. These morphological data are supported in part by investigations which indicate that norepinephrine may play a role in the secretion of VP from the posterior pituitary *(Kuhn, 1974; Milton and Paterson, 1974)*. A method recently developed by this laboratory, for the simultaneous visualization of monoamines and neuropeptides in a single tissue block *(McNeill and Sladek 1978; Sladek et al., 1978c)* has been used to demonstrate an apparent morphological juxtaposition between noradrenergic terminals and OX and VP- containing soma of the SON and PVN *(McNeill and Sladek, 1977)*. This technique provides a means for the simultaneous analysis of alterations in these systems at a specific point in time (e.g. ontogeny, puberty, aging) or during selected functional states (e.g. osmotic stress, reproduction). Since existing data indicate that the regulation of water balance may be altered significantly in old age *(Rodeck, Lederis and Heller, 1960; Donihue, 1965; Turkington and Everitt, 1976)*, the present investigation was undertaken to examine the integrated morphology of both CA and neurophysin systems in young and old macaques in an attempt to provide a morphological correlate to this phenomenon.

MATERIALS AND METHODS

Six macaques *(M. nemestrina)*, three each of 4 and 20 years of age, were killed under ketamine anesthesia (10 mg/kg i.m.). These animals were maintained under monitored conditions at the Seattle Regional Primate Center and appeared generally free from disease and malnutrition. Three pairs of animals (one of each age per pair) were freeze-dried as a unit for 8 weeks and treated for Falck-Hillarp histochemistry as described previously by this laboratory *(Hoffman and Sladek, 1973)*. Each set of three brains was embedded into a single paraffin block and serially sectioned at 10 μM as a single specimen. Every tenth section was stained with cresyl violet for orientation purposes. Anatomically similar levels through the PVN were chosen and examined for the simultaneous demonstration of neurophysin and CA. Immunocytochemistry of neurophysin was performed on alternate sections to those examined for histofluorescence according to the method of McNeill and Sladek, *(1978)*. Bovine neurophysin (BNP), human estrogen-stimulated neurophysin (ESN) and human nicotine-stimulated neurophysin (NSN) antisera were supplied by E. A. Zimmerman and A. G. Robinson. BNP was used to analyze all neurophysin-containing cells while the more specific ESN and NSN antisera demonstrated OX and VP-synthesizing neurons respectively *(Robinson, 1975; Robinson, 1976)*. Examination was made of the qualitative appearance and relative number of neurophysin-containing perikarya in each experimental pair. Simultaneous visualization of CA varicosities and neurophysin-containing perikarya of the PVN was performed to determine the relative degree to which CA-neurophysin interactions occur in young and old macaques.

RESULTS

Examination of sections stained with cresyl violet revealed that the PVN of *Macaca nemestrina* appeared similar in position and shape to the human PVN. Neurons within the PVN were generally multipolar although some bipolar ones were present. Processes of these cells were seldom seen in Nissl preparations, but were visualized in sections stained for BNP. Quantitative cell counts of the magnocellular perikarya in the PVN revealed a comparable total number of neurons present in the young and old monkey.

Immunocytochemical staining of the PVN with BNP revealed the presence of two populations of stained neurons, one lightly stained and the other densely stained. Both populations of neurons possessed multipolar shapes and processes filled with stained neurophysin which extended 20-100 μm from the cell perikaryon. This dual population of neurons appeared in both young and old macaques, but a difference in the relative numbers of each population was noted as described below. Furthermore, the density of staining in the dark cell population appeared to decrease with age. The 20-year old monkeys were characterized by the presence of large, densely stained Herring bodies within both the PVN and the exiting fibers of the hypothalamoneurohypophyseal tract. The total amount of immunoreactive product appeared reduced in the old animals in part due to a reduced staining of neuronal processes.

CATECHOLAMINE HISTOFLUORESCENCE

Dense patterns of CA varicosities were seen within the hypothalamus of young and old macaques. In general, the density of varicosities within identified hypothalamic nuclei did not appear to change with age although a slight increase was noted within the arcuate nucleus wherein patterns approached high (4+) levels in the older specimens. The magnocellular nuclei were characterized by extremely dense patterns of CA varicosities, many of which appeared in juxtaposition to the magnocellular perikarya as described below.

QUANTITATIVE NEUROPHYSIN ANALYSIS

Quantitation of the number of neurons stained with BNP neurophysin revealed no change in the total cell population of the PVN with age; however, independent analysis of dark and light stained cells of the PVN revealed a marked reduction in the dark stained cells with a concommitant increase in light stained cells. Cell counts performed according to the techniques of Konigsmark *(1970)* revealed an 84% decrease in dark-stained cells in the PVN in the older monkey. Staining with the more specific ESN and NSN antisera confirmed the finding of a constant total number of immunoreactive cells. Of interest, dark stained cells with ESN antiserum did not demonstrate a change in number while those stained with NSN

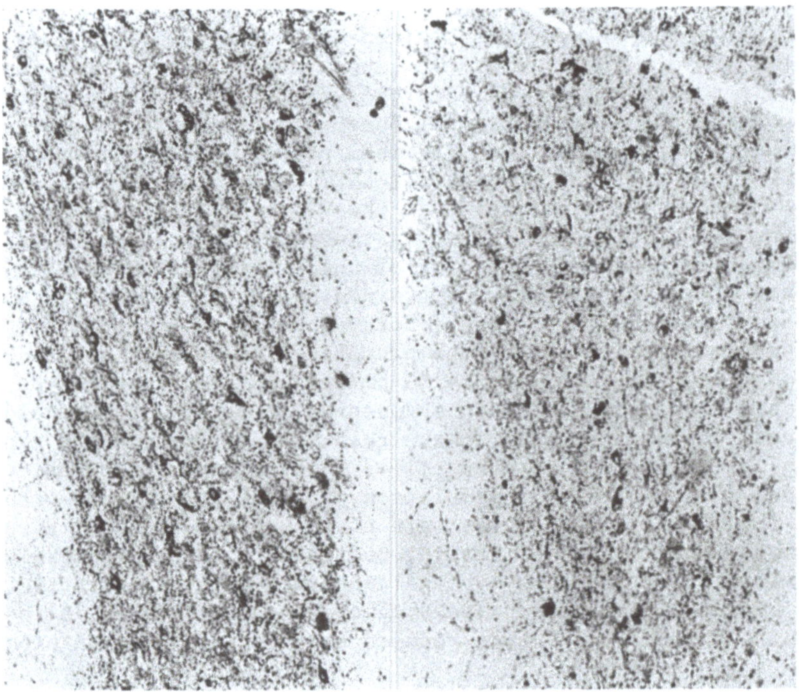

Figure 1. Left. Paraventricular nucleus: 4 year old M. nemestrina. Immunocytochemical staining for NSN demonstrates numerous dense neuronal perikarya and an abundance of beaded processes. X60

Figure 2. Right. Paraventricular nucleus: 20 year old M. nemestrina. A neuronanatomical level comparable to that illustrated in Fig. 1 demonstrates a reduction in NSN positive parikarya and processes. X60.

decreased with age by 54% (Figures 1 and 2). In the latter instance, a concommitant rise in light stained cells was seen. Immunocytochemical staining with the specific ESN and NSN antisera revealed a marked drop in positively stained fibers of the hypothalamo-neurohypophyseal tract and within the PVN. Large positively-stained Herring bodies were present with both antisera.

CATECHOLAMINE-NEUROPHYSIN INTEGRATION

Analysis of alternate tissue sections with the simultaneous visualization technique for neuropeptides and monoamines was applied to separate sections stained with each neurophysin antiserum. In all instances, CA varicosities were seen in juxtaposition to neurophysin-containing perikarya and proximal dendrites. Numerous CA varicosities appeared to align with the perimeter of these neurons (Figures 3 and 4), although the majority of CA varicosities did not appear in juxtaposition to neuronal perikarya. Analysis of the percentage of cells which appeared contacted by CA varicosities in ESN and NSN stained sections revealed a certain constancy in young and old animals. However, a slight decrease in the number of ESN-dark stained cells was noted between young and old animals (67% young; 58% old). Also, a slight increase was noted in the number of NSN-light stained cells between the 4 and 20 year olds (70% young; 76% old) which appeared contacted by CA varicosities.

DISCUSSION

The use of an immunocytochemical approach to evaluate age-related changes in neuropeptide systems presents a more sensitive and selective method to study peptide morphology than has been employed previously. Since BNP is non-selective for a specific neurophysin in the monkey, both populations of neurophysins are stained when BNP is used as the primary antiserum *(Zimmerman, et al., 1975)*. The presence of a dual population of immunoreactive cells (i.e. dark and light staining neurons) in the magnocellular nuclei stained with BNP has been reported previously in both the adult rhesus monkey *(Zimmerman, 1973)* and rat *(Sokol, Zimmerman, Sawyer and Robinson, 1976)*. This difference in the density of cytoplasmic staining between neurons may not solely represent a difference in peptide content, but may also reflect a somewhat selective yet not totally specific preference for one of the monkey neurophysins as has been reported in the rat *(Sokol et al., 1976)*. Therefore, cautious interpretation of comparisons between changes in light and dark cells is imperative. The present data showed a marked reduction in the subpopulation of dark staining neurons with BNP in the 20 year-old animal. Although densitometry measurements were not performed, it was apparent that both light and dark cell populations of the PVN in the aged animal were somewhat less dense than in the young animal. Neuronal

Figure 3. *Upper paraventricular nucleus: 4-year old M. nemestrina. Catecholamine varicosities and neurophysin (ESN) perikarya are seen simultaneously. Numerous varicosities (→) appear to surround this peptidergic neuron of the magnocellular system. X370.*

Figure 4. *Lower paraventricular nucleus: 20-year old M. nemestrina. Catecholamine varicosities (→), as in the younger animals, appear in juxtaposition to a peptidergic neuron. The cell perikaryon and a thick process appear to be contacted. X370.*

perikarya and axonal and dendritic processes appeared to contain less reaction product in the older animals. This reduction in number of dark-stained cells as well as the overall decrease in the amount of reaction product in the PVN suggests that there is a decrease in the amount of peptide and carrier content of the PVN in the 20-year old animal. Since the primary site of magnocellular neuropeptide storage is in axon terminals in the neurohypophysis and not the perikarya of the PVN and SON *(Sloper, 1966)*, this alteration in peptide content may represent a depressed synthetic capability in the older monkey as opposed to a storage phenomenon. However, whether this dysfunction results from an intrinsic malfunction of the synthetic machinery, an alteration in the afferent input or some other cellular mechanism must await further investigation.

Staining with the specific ESN and NSN antisera provided further information as to specific cell population changes within the PVN in the old monkey. Using ESN as the primary antiserum, there was no change in the population of dark cells in the PVN of the young and old monkeys, however, a reduction of 54% was noted in dark stained cells with the NSN antiserum. These data suggest the possibility of a preferential effect on VP-synthesizing neurons as opposed to OX-synthesizing neurons. It is important to note, however, that the percentage drop in NSN stained cells was not as great as that with BNP staining. Presumably this is accounted for by a certain degree of cross-reactivity of the NSN antiserum with OX neurons, since both neurophysins may contain a number of homogeneous antigenic peptide sequences *(Capra and Walter, 1975)*. The presence of an increased number of large Herring bodies in the 20-year old monkey may also be implicated in age-related changes in water balance. Since the presence of Herring Bodies may result from an alteration or disruption on an axoplasmic flow *(Dellman, 1973)*, the presence of a substantial increase in both the size and number of these inclusions in the old monkey is significant and suggests that there may be a substantial change in axoplasmic flow to the neural lobe accompanying old age. This finding supports earlier findings of a decreased amount of neurosecretory material in the posterior pituitary in old animals using both the Gomori stain *(Rodeck et al., 1960; Donihue, 1965)*

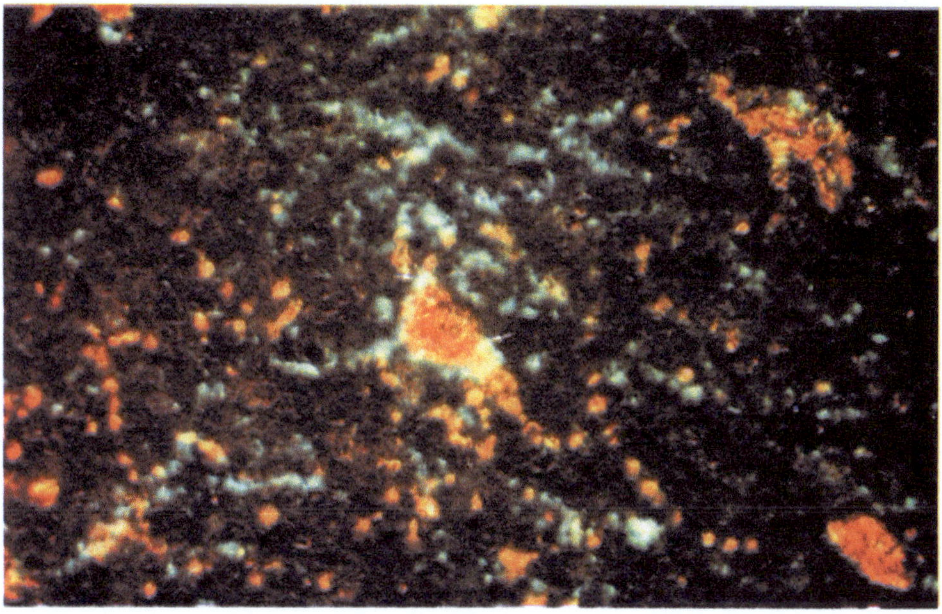

Fig. 3

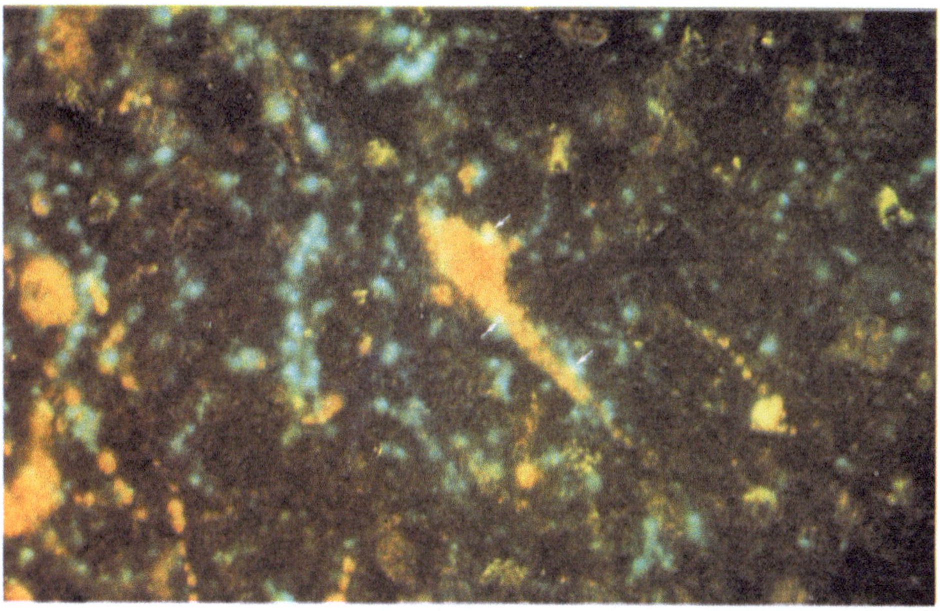

Fig. 4

and bioassay *(Turkington and Everitt, 1976)*. These data suggest that the decrease of neurosecretory material in the posterior pituitary of the old mammal may result from an alteration in axoplasmic flow rather than an alteration in intrinsic storage or secretory capabilities of the endocrine gland.

Simultaneous visualization of CA varicosities and neurophysin-containing perikarya revealed a high percentage of apparent axosomatic juxtaposition in PVN in both the young and old animals with both ESN and NSN antisera. This is a departure from the phenomenon reported in rat brain by Fuxe *(1965)* and Sladek, McNeill and Zimmerman *(1978)* wherein magnocellular PVN perikarya appeared somewhat ventrolateral to the CA field. This finding may represent a high degree of CA influence on PVN perikarya in the primate, an influence which could persist with aging based on the morphology presented here. However, these findings do not account for all axodendritic terminations, which could provide an important input into this magnocellular nucleus. Such analysis must await a methodology which allows visualization of more than just the proximal portion of dendrites as seen with the present immunocytochemical technique.

Qualitative histofluorescence analysis did not reveal a striking change in CA terminal innervation of this nucleus. However, the subjective observation was that of a possible shift in density of fine terminals which could be candidates for axodendritic terminations. These data appear to be consistent with a concurrent report by Sladek, McNeill, Walker and Sladek *(1978)* of a lack of qualitative histofluorescence change in the ventral noradrenergic bundle in these same animals. This is in contrast to a significant reduction in intraneuronal content of the perikarya of origin of the dorsal noradrenergic bundle. The ventral bundle has been shown in rat *(Ungerstedt, 1971)* to innervate the hypothalamus and might be presumed in primate to act in a similar manner. The observed change in fine terminal innervation of PVN, however, could be attributed to one or two additional sources of afferent input to this nucleus, one being from the dorsal bundle itself or another from a previously undescribed dopaminergic input to this nucleus. Unpublished observations in our laboratory indicate that, at least in the rodent, a dopaminergic input may exist to the PVN; although no evidence at present indicates this may occur in the primate, it remains a viable possibility, particularly in light of the marked drop in dopaminergic neurons in primate brain stem as reported elsewhere in this proceedings *(Sladek and Sladek, 1978)*. The former feasibility of a coeruleal type innervation from the dorsal noradrenergic bundle is especially interesting considering the possibility that the dorsal bundle might contribute fibers to the hypothalamus of the macaque *(Bowden, German and Poynter, 1978)*.

PVN perikarya of the older animals demonstrated a relatively higher intraneuronal content of lipofuscin than in younger animals. The presence of lipofuscin in coeruleal neurons was demonstrated to be coincidental with a decreased content of norepinephrine as determined by Sladek and Sladek *(1978)*. If a similar phenomenon occurs in neurons which contain peptides and lipofuscin then one might account for the generalized decreased density of staining of neurophysin in the older animals based on the presence of this autofluorescent pigment.

ACKNOWLEDGEMENTS

Supported by USPHS Grant NS 11642 (JRS). We thank Betsy Putnam Buell for technical assistance and Douglas M. Bowden, Seattle Regional Primate Center for prividing the primate tissue for this study.

REFERENCES

Bowden, D.M., German, D.C. and Poynter, W.D. (1978). An autoradiographic, semistereotaxic mapping of major projections from locus coeruleus and adjacent nuclei in *Macaca mulatta*. *Brain Res. 151*, (in press).

Capra, J.D. and Walter, R. (1975). Primary structure and evolution of neurophysins. *Ann. N.Y. Acad. Sci. 248*, 397-407.

Dellmann, H.D. (1973). Degeneration and regeneration of neurosecretory systems. *Int. Rev. Cytol. 36*, 215-315.

Donihue, F.W. (1965). Reduced juxtaglomerular cell granularity, pituitary neurosecretory material and width of the zona glomerulosa in aging rats. *Endocrinology 77*, 948-951.

Fuxe, K. (1965). Evidence for the existence of monoamine neurons in the central nervous system, IV The distribution of monoamine nerve terminals in the central nervous system. *Acta physiol. scand., suppl. 64*, 247.

Hoffman, D.L. and Sladek, J.R., Jr. (1973). The distribution of catecholamines within the inferior olivary complex of the gerbil and rabbit. *J. comp. Neurol. 151*, 101-112.

Hokfelt, T., Elde, R., Fuxe, K., Johansson, O., Ljungdahl, Ake, Goldstein, M., Luft, R., Efendic, S., Nilsson, G., Terenius, L., Ganten, D., Jeffcoate, S.L., Rehfeld, J., Said, S., Perez de la Mora, M., Possani, L., Tapia, R., Teran, L. and Palacios, R. (1978). Aminergic and peptidergic pathways in the nervous system with special reference to the hypothalamus. In: *The Hypothalamus, Research Publications: Assiciation for Research in Nervous and Mental Disease, vol. 56*, pp. 69-135. Eds. S. Reichlin, R.J. Baldessarini and J.B. Martin, Raven Press, New York

Konigsmark, B.W. (1970). Methods for the counting of neurons. In: *Contemporary Research Methods in Neuroanatomy*, pp. 315-340. Eds. W.J.H. Nauta and S.O.E. Ebbesson, Springer Verlag, N.Y.

Kuhn, E.R. (1974). Cholinergic and adrenergic release mechanism for vasopressin in the male rat: A study with injections of of neurotransmitters and blocking agents into the third ventricle. *Neuroendocrinology 16*, 255-264.

McNeill, T.H. and Sladek, J.R., Jr., (1977). Correlative fluorescence-immunocytochemical localization of monoamines and neurophysin, vasopressin and gonadotropin-releasing hormone in the rat and monkey hypothalamus. *Proc. Soc. Neurosci. 3*, 351.

McNeill, T.H. and Sladek, J.R., Jr., (1978). Fluorescence-immunocytochemistry: simultaneous localization of catecholamines and gonadotropin-releasing hormone. *Science 200*, 72-74.

Milton, A.S. and Paterson, A.T. (1974). A microinjection study of the control of antidiuretic hormone release by the supraoptic nucleus of the hypothalamus in the cat. *J. Physiol. 241*, 607-628.

Robinson, A.G. (1975). Isolation, assay, and secretion of individual human neurophysins. *J. Clin. Invest. 55*, 360-367.

Robinson, A.G., Ferin, M. and Zimmerman, E.A. (1976). Plasma neurophysin levels in monkeys: emphasis on the hypothalamic response to estrogen and ovarian events. *Endocrinology 98*, 468-475.

Rodeck, H., Lederis, K. and Heller, H. (1960). The hypothalamo-neurohypophysial system in old rats. *J. Endocr. 21*, 225.

Sladek, J.R., Jr., McNeill, T.H. and Zimmerman, E.A. (1978). Monoamine-neurophysin interrelationships. *Anat. Rec. 190*, 544.

Sladek, J.R., Jr., McNeill, T.H., Walker, P. and Sladek, C.D. (1978). Age-related alterations in monoamine and neurophysin systems in primate brain. In: *Aging in non-human primates, Neurobiology of Aging, vol. 6.* Ed. D.M. Bowden, Raven Press, New York, (in press).

Sladek, J.R., Jr., Sladek, C.D., McNeill, T.H. and Wood, J.G. (1978). New sites of monoamine localization in the endocrine hypothalamus as revealed by new methodological approaches. In: *Neural Hormones and Reproduction*, pp. 154-171. Eds. D.E. Scott, G.P. Kozlowski and A. Weindl, Karger, Basel.

Sladek, J.R., Jr. and Sladek, C.D. (1978). Relative quantitation of monoamine histofluorescence in young and old non-human primates. In: *Parkinson's Disease: Aging and Neuroendocrine Relationships*, Eds. C.E. Finch, D.E. Potter and A.D. Kenny, Plenum Press, N.Y.

Sloper, J.C. (1966). The experimental and cytopathological investigation of neurosecretion in the hypothalamus and pituitary. In: *The Pituitary Gland*, pp. 131-239. Eds. G.W. Harris and B.T. Conovan, University of California Press, Berkeley.

Sokol, H.W., Zimmerman, E.A. Sawyer, W.H. and Ribinson, A.G. (1976). The hypothalamic-neurohypophyseal system of the rat: localization and quantitation of neurophysin by light microscopic immunocytochemistry in normal rats and in Brattleboro rats deficient in vasopressin and a neurophysin. *Endocrinology 98*, 1176-1188.

Turkington, M.R. and Everitt, A.V. (1976). The neurohypophysis and aging with special reference to the antidiuretic hormone. In: *Hypothalamus, Pituitary and Aging*, pp. 123-136. Eds. A.V. Everitt and J.A. Burgess, Charles C. Thomas Publishing, Springfield.

Ungerstedt, U. (1971). Stereotaxic mapping of monoamine pathways in the rat brain. *Acta physiol. scand. suppl. 367,* 1-48.

Zimmerman, E.A., Hsu, K.C., Robinson, A.G., Carmel, P.W., Frantz, A.G. and Tannenbaum, M. (1973). Studies of neurophysin-secreting neurons with immunoperoxidase techniques employing antibody to bovine neurophysin. I. Light microscopic findings in monkey and bovine tissues. *Endocrinology 92,* 931-940.

Zimmerman, E.A., Defendini, R., Sokol, H.W. and Robinson, A.G. (1975). The distribution of neurophysin-secreting pathways in the mammalian brain: light microcopic studies using the immunoperoxidase technique. *Ann. N.Y. Acad. Sci. 248,* 92-111.

LOSS OF CHOLINE ACETYLTRANSFERASE ACTIVITY IN NORMAL AGING AND IN SENILE DEMENTIA

PETER DAVIES

Department of Pathology
Albert Einstein College of Medicine
1300 Morris Park Avenue
Bronx, New York 10461

INTRODUCTION

The goal of neurochemical studies on human brain is to find out how various sub-populations of neurons are affected by aging and disease. Neurons are classified according to the substance used for chemical transmission, and the groups can then be investigated by measuring the activities of enzymes specifically involved in the formation or degradation of the transmitter. Neurons which use acetylcholine as the transmitter are among the easiest to investigate in human brain, because the enzyme choline acetyltransferase is found uniquely in such cells *(Kuhar, 1976)*. There is also abundant evidence to show that the enzyme activity is stable in the brain for periods of days following death, a feature which greatly facilitates studies of autopsied human tissues *(reviewed by MacKay, Davies, Dewar and Yates, 1978)*.

My own studies over the last few years have concentrated on the effects of aging and dementia on the human cholinergic neurons of the cerebral cortex. Before discussing the results of these studies, I would like to try to outline the main problems encountered by myself and others in research of this type. These difficulties center around the selection of tissues for neurochemical examination.

CASE SELECTION

Definition of a control group of subjects is not nearly as simple as one would wish. To accurately determine the influence of age alone on human cholinergic neurons, the ideal would be to obtain brains from individuals of various ages in whom the level of cognitive function was not altered significantly by illness in the pre-terminal phase. In practical terms, the best I have been able to achieve is to determine whether or not individuals coming to autopsy were alert, responsive, and orientated with respect to time and place within a few hours of death. Where this could be determined, the case notes were examined in more detail for indications of any psychiatric or neurological illness. If this screening revealed nothing abnormal, the case was tentatively labelled normal, and the right hemisphere was taken for neurochemical investigation. The left hemisphere, the brain stem, and the cerebellum were fixed in formalin and sent for neuropathological examination. Any of the following abnormalities was considered sufficient reason to exclude the case: gross depigmentation of the substantia nigra or locus coeruleus; anything more than minimal cerebral atherosclerosis; presence of old or new infarcts; neurofibrillary tangles or senile plaques.

This clinical and neuropathological screening process has eliminated about 80% of the cases, without prior diagnosis of dementia, coming to autopsy over the last year. Even with this selection it is hard to know how accurately the control group represents the population at large. The definition of a group of cases of senile dementia of the Alzheimer type (SDAT) presents rather less of a problem. Almost all cases with a diagnosis of SDAT are thoroughly evaluated by psychiatrists and neurologists before death, and so the mental status is usually well documented. In my studies, the clinical notes have been evaluated to choose the cases with profound, progressive global dementia, with no focal neurological signs. Those cases chosen almost always showed numerous neurofibrillary tangles and neuritic plaques in sections of cerebral cortex or neuropathological study. Cases also showing significant cerebrovascular disease or depigmentation of either substantia nigra or locus coeruleus were eliminated so that a clear picture of SDAT could be obtained. Again for the purposes of clarity, SDAT cases were only included in the study if numerous neurofibrillary tangles and neuritic plaques were found in sections of hippocampus *and* sections of frontal and parietal cortex.

Although the procedures for case selection outlined above have been very tedious and time-consuming, over the last four years the data gathered on the cortical cholinergic system have been reasonably clearcut.

RESULTS AND DISCUSSION

The first and most obvious conclusion to emerge from studies of choline acetyltransferase (ChAT) activity in the cases selected was the large reductions in several areas of cerebral cortex from SDAT cases. Some of these data are presented in Table I. There are two reasons for dividing the cases, both controls and SDAT, into those aged 70 or less and those aged 71 or more.

Soon after Maloney and Davies *(1976)* first reported reductions in ChAT actively in SDAT cases, publications from two inde-

TABLE 1. CHOLINE ACETYLTRANSFERASE ACTIVITY IN FOUR BRAIN REGIONS

(A) CONTROLS AND SDAT CASES 70 YEARS OF AGE OR LESS AT DEATH

AREA	CONTROL	SDAT
Convexity Frontal Cortex	0.47 ± 0.01 (13)	0.030 ± 0.003 (6)
Parietal Cortex	0.36 ± 0.05 (13)	0.017 ± 0.003 (6)
Mid-Temporal Gyrus	0.45 ± 0.04 (13)	0.019 ± 0.007 (6)
Hippocampus	0.53 ± 0.05 (13)	0.033 ± 0.003 (6)

Activities of the SDAT group are significantly lower than those of the controls ($p < 0.01$).

(B) CONTROLS AND SDAT CASES 71 YEARS OF AGE OR MORE AT DEATH

AREA	CONTROL	SDAT
Convexity Frontal Cortex	0.05 ± 0.005 (7)	0.02 ± 0.005 (8)
Parietal Cortex	0.04 ± 0.008 (7)	0.01 ± 0.004 (8)
Mid-Temporal Gyrus	0.04 ± 0.006 (7)	0.02 ± 0.005 (8)
Hippocampus	0.16 ± 0.020 (7)	0.05 ± 0.012 (8)

Activities of the SDAT group are significantly lower than those of the controls ($p < 0.05$).

SDAT: senile dementia of Alzheimer type

pendent groups *(Perry, Perry, Blessed and Tomlinson, 1977; White, Goodhardt, Keet, Hiley, Carasco, Williams and Bowen, 1977)* confirmed our findings. Notable, however, were the apparently smaller reductions in ChAT activity found by these workers when their data and ours were compared. Detailed scrutiny of the papers cited, as well as later work *(Perry, Gibson, Blessed, Perry and Tomlinson, 1977)*, revealed that Perry and her co-workers had selected cases appreciably older in mean age than those I had studied, and that this was true also of the study of White and colleagues.

As Table I shows, the extent of the difference in ChAT activity between control and SDAT cases is dependent largely on the age range of the control group. This is because the older controls have a much lower cortical Chat activity than the younger controls. There are not significant differences in ChAT activity between the younger and older groups of SDAT cases. This point is better displayed in Figures 1 and 2, which show the actual ChAT activities found in two brain regions: in the control cases there is a clear

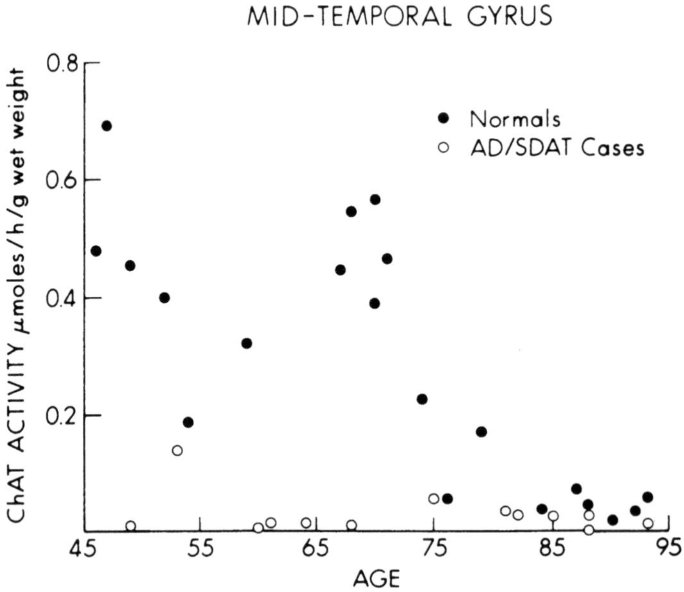

Figure 1. The decline in ChAT activity with age in the normals is clearly seen. The activity in the SDAT cases is uniformly low.

age-related decline, which is statistically significant in both regions (mid-temporal gyrus, r= 0.773, t= 5.17; hippocampus, r= 0.847, t= 6.56).

The second reason for separating the data from younger and older SDAT cases relates to the long-standing confusion regarding the nomenclature of this condition. Almost by tradition, demented patients below the age of 65 who were shown to have neurofibrillary tangles and neuritic plaques were said to have Alzheimer's disease, or pre-senile dementia. Those aged 65 or older with similar pathology were usually labelled senile dementia cases. As Katzman *(1976)* has pointed out, the clinical and pathological features of the under 65 and over 65 cases are virtually identical. The data presented in this paper suggest that the pre-senile and senile forms of Alzheimer's disease are remarkably similar neurochemically, and thus support the view that the apparently arbitrary age of 65 should no longer be used to separate patients suffering from what seems to be a single disease entity.

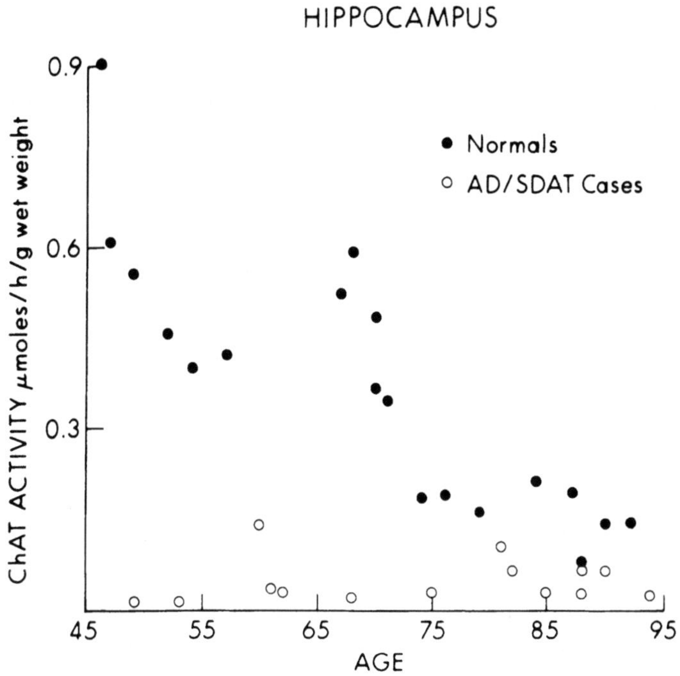

Figure 2. A strong inverse correlation between age and ChAT activity is again evident in the normals. The activity in the SDAT cases is uniformly low.

CONCLUSIONS

That there are major deficits in ChAT activity in the cerebral cortex of SDAT cases now seems beyond doubt. What these data tell us about the disease is not yet clear. Future research must distinguish between extensive loss of cholinergic neurons in SDAT, and/or survival of these cells with greatly reduced functional capacity. This question has enormous potential significance. Muscarinic receptors for acetylcholine are present in normal numbers in SDAT cortex *(Davies and Verth, 1977)*, and if the cholinergic neurons do survive the disease, then steps to restore their function may have significant therapeutic benefits.

The age-related decline in ChAT activity found in the control cases suggests that cholinergic function declines in the absence of brain disease. Drachman and Leavitt *(1974)* came to a similar conclusion as a result of their elegant pharmacological studies with young and old people. It may well be that neurochemical and pharmacological studies can now explain, at least in part, a very common observation: that aging is associated with considerable loss of mental faculties.

REFERENCES

Davies, P. and Maloney, A.J.F. (1976). Selective loss of central cholinergic neurons in Alzheimer's disease. *Lancet ii*, 1403.

Davies, P. and Verth, A.H. (1977). Regional distribution of muscarinic acetylcholine receptor in normal and Alzheimer's-type dementia brains. *Brain Research 138*, 385-392.

Drachman, D.A. and Leavitt, J. (1974). Human memory and the cholinergic system. *Arch. Neurol. 30*, 113-121.

Katzman, R. (1976). The prevalence and malignancy of Alzheimer's disease: a major killer. *Arch. Neurol. 33*, 217-218.

Kuhar, M.H. (1976). The anatomy of cholinergic neurons. *In:* A. M. Goldberg and I. Hanin (Eds.) *Biology of Cholinergic Function.* Raven Press, New York, pp 3-28.

MacKay, A.V.P., Davies, P., Dewar, A.J. and Yates, C.M. (1978). Regional distribution of enzymes associated with neurotransmission by monoamines, acetylcholine and GABA in the human brain. *J. Neurochem. 30*, 827-839.

Perry, E.K., Gibson, P.H., Blessed, G., Perry, R.H. and Tomlinson, B.E. (1977). Neurotransmitter enzyme abnormalities in senile dementia. *J. Neurol. Sci. 34*, 247-265.

Perry, E.K., Perry, R.H., Blessed, G. and Tomlinson, B.E. (1977). Necropsy evidence of central cholinergic deficits in senile dementia. *Lancet i*, 189.

White, P., Goodhardt, M.J., Keet, J.P., Hiley, C.R., Carasco, L.H., Williams, I.E.I. and Bowen, D.M. (1977). Neocortical neurons in elderly people. *Lancet i*, 668-671.

SUBJECT INDEX

MIX
Papier aus verantwortungsvollen Quellen
Paper from responsible sources
FSC® C105338

If you have any concerns about our products,
you can contact us on
ProductSafety@springernature.com

In case Publisher is established outside the EU,
the EU authorized representative is:
Springer Nature Customer Service Center GmbH
Europaplatz 3, 69115 Heidelberg, Germany

Printed by Libri Plureos GmbH
in Hamburg, Germany